Information, Communication and Education on Climate Change –
European Perspectives

# Environmental Education, Communication and Sustainability

**edited by Walter Leal Filho**

**Vol. 26**

PETER LANG
Frankfurt am Main · Berlin · Bern · Bruxelles · New York · Oxford · Wien

Walter Leal Filho
Franziska Mannke
Philipp Schmidt-Thomé
(eds.)

# Information, Communication and Education on Climate Change – European Perspectives

PETER LANG
Internationaler Verlag der Wissenschaften

**Bibliographic Information published by the Deutsche Nationalbibliothek**
The Deutsche Nationalbibliothek lists this publication in the Deutsche Nationalbibliografie; detailed bibliographic data is available in the internet at <http://www.d-nb.de>.



AstrA

Baltic Sea Region
INTERREG III B

Project part-financed by the European Union
(European Regional Development Fund)
within the BSR INTERREG III B NP programme

ISSN 1434-3819
ISBN 978-3-631-56682-4



Printed in Germany 1 2 3 4 6 7

www.peterlang.de

## Table of Contents

# Preface

It is now beyond any doubt that climate change represents a major threat to the environmental, social and economic well being of the planet.

During the last century, the Earth's average surface temperature rose by around 0.6°C and has not stopped since. Current evidence is becoming stronger that most of the global warming that has occurred over the last 50 years is at least partly attributable to human activities. In its Forth Assessment Report, published in early 2007, the Intergovernmental Panel on Climate Change (IPCC) projects that global average surface temperatures will rise further by the end of this century. It is a fact that such global temperature increase is likely to lead to serious implications for both human beings and other life forms. It may also lead to a rise in sea levels of an estimated 9 to 88 cm by the end of this century, which will endanger coastal areas and small islands, and a greater frequency and severity of extreme weather events.

It is widely acknowledged that there are many human activities which contribute to climate change. These include on the one hand the burning of fossil fuels and, on the other hand, deforestation. Both phenomena lead to higher emissions of carbon dioxide (CO2), which is the main gas responsible for climate change. But they also produce other 'greenhouse' gases. Therefore, if we want to bring climate change to a halt, global greenhouse gas emissions must be reduced significantly.

There are now many mechanisms in place to address the challenges posed by climate change. The United Nations Framework Convention on Climate Change and its Kyoto Protocol are examples of what is being done. However, more is needed. A further means to work towards a reduction in greenhouse emissions is by raising awareness on the causes and consequences of climate change. Only by motivating and engaging people to actively consider how their habits and behaviours contribute towards the problem, can we expect it to be solved in the long-term.

This book is an attempt to contribute to the global debate on climate change by outlining some educational methods, approaches and projects which have been focusing on climate issues in Europe. The examples provided in the book illustrate a point often made but unfortunately not sufficiently pursued: education, communication and training can play a key role in supporting efforts towards addressing climate change.

We hope this book and its contents will convince those which are still sceptical about the role education can play in dealing with climate change and that it may catalyse further educational initiatives worldwide.

*Walter Leal Filho, Franziska Mannke, Philipp Schmidt-Thomé* Summer 2007

Chapter 1

# ESPACE
# –
# A New Transnational Approach to Adapting to the Impacts of Climate Change

**Jill Cook[1], Doogie Black[1], Angela Bentham[1], Tania Stadsbader[2], Jorn Peters, Lesley van Dijk[3], Mark Goldthorpe[4], Tim Yair, Fran Wallington[5], Meinte de Hoogh[6], Michael Belau[7]**

## Abstract

However successful we are at reducing emissions, we expect to deal with many decades of climate change due to emissions that we have already put into the climate system. We need to start adapting today to these changing conditions to ensure our social, economic and environmental systems are best-prepared for the unavoidable risks and are able to take advantage of any opportunities. Although spatial planning will play a crucial part in helping communities adapt to climate change, our current planning systems do not sufficiently take climate change into account and are therefore not robust enough to deal with the impacts. This paper summarises the work of ESPACE (European Spatial Planning: Adapting to Climate Events) – an Interreg IIIB North West Europe project which aims to tackle this important issue by recommending how adaptation to climate change can be incorporated into the spatial planning system.

## Introduction

Europe is already experiencing the effects of climate change. The devastating heatwave of 2003 for example, that lead to around 35,000 deaths across Europe (Bhattacharya 2003) has now been linked to climate change (Stott et al. 2004). Temperatures in France soared to 104 degrees Fahrenheit (40 degrees Celsius) and remained unusually high for two weeks. By the 2040s, temperatures as high as this

---

1 Hampshire County Council.
2 Regionaal Landschap Zenne, Zuun and Zoniën.
3 South East England Regional Assembly.
4 South East Climate Change Partnership.
5 West Sussex County Council.
6 Ministerie van VROM.
7 Bayerisches Landesamt für Umwelt.

are expected to occur every other year. By the 2060s, summers like 2003 will typically be coolest in the decade (Met Office 2005). This predicted increase in the risk of heatwaves, together with greater frequency of extreme weather events, hotter drier summers and wetter winters has put climate change firmly on the agenda across Europe. Scientists, policy makers and politicians are all in agreement that this is one of the biggest challenges facing Europe today.

Mitigation measures to reduce greenhouse gas emissions are essential if we are to avoid the worst effects of climate change. Yet, however successful we are at reducing emissions, we expect to have to deal with many decades of climate change due to emissions that we have already put into the climate system. We need to start adapting today to these changing conditions, to ensure our social, economic and environmental systems are best-prepared for the unavoidable risks and are able to take advantage of any opportunities. The UK Stern Review on the Economics of climate change highlights that we now know enough about the issue to assess and plan for the risks (Stern 2006). Spatial planning[1] was identified in this review as one of four fundamental areas that will play a crucial part in helping communities 'adapt' to climate change. However, our current planning systems do not sufficiently take climate change into account and are therefore not robust enough to deal with the impacts. Problems with the current system include the fact that:

- Climate change is not currently given the weight it deserves in spatial planning
- The consequences of a planning decision last longer than the spatial planning period. For example in the UK, built development might be expected to last 80 years whereas the spatial plan only lasts 25 years. This is not an issue if set against a backdrop of a stable climate. However *with* climate change, spatial planners must take account of changing climate risks that will affect the whole of the decision 'lifetime' rather than just the spatial plan period.
- We face some difficult decisions now and in the future in managing the impacts of climate change, particularly for existing developments. Some developments in particularly vulnerable locations may prove unsustainable in the longer-term, in the face of changing climate risks.

These are just some of the key issues being addressed by ESPACE – a groundbreaking European project whose final results are due to be launched in the summer of 2007.

---

1 Spatial planning is „a process that assimilates and interprets evidence-based knowledge to inform those activities that aim to ensure spatial development takes place in an appropriate way, from a functional, social, economic and environmental point of view." Working definition of the ESPACE Project 2007.

## What is ESPACE?

ESPACE (European Spatial Planning: Adapting to Climate Events), is a four year project led by Hampshire County Council and funded by the European Commission's North West Europe INTERREG IIIB Programme, the ESPACE Partnership and the Department for Communities and Local Government. ESPACE aims to influence the philosophy and practice of spatial planning by recommending how adaptation to climate change can be incorporated. Concentrating on water management issues, it was one of the first projects of its kind to focus on increasing awareness of the need for spatial planning systems to adapt to the impacts of climate change and to begin to provide some of the necessary policy guidance, tools and mechanisms to incorporate adaptation into planning systems and processes.

Fig. 1: ESPACE is developing solutions for better water resource management, including water storage options. Credit – Ministerie van VROM

Recognising that climate change is a global issue transcending political boundaries, ESPACE has been founded by a transnational group of 10 partners, also known as the 'ESPACE Partnership', spanning four North West European countries and bringing together representatives from all levels of civic society. They include, *Hampshire County Council (Lead Partner)*, *the Environment Agency*, *South East Climate Change Partnership*, *South East England Regional Assembly*, *Surrey County Council and West Sussex County Council* from the UK, *Regionaal Landschap Zenne, Zuun en Zoniën* from Belgium, *Waterschap Rivierenland and*

*Ministerie van VROM* from the Netherlands and the *Bayerisches Landesamt für Umwelt* from Germany.

Fig. 2: The ESPACE Partnership is made up of 10 organisaitons spanning four North West European countries. Credit – Interreg IIIB North-West Europe

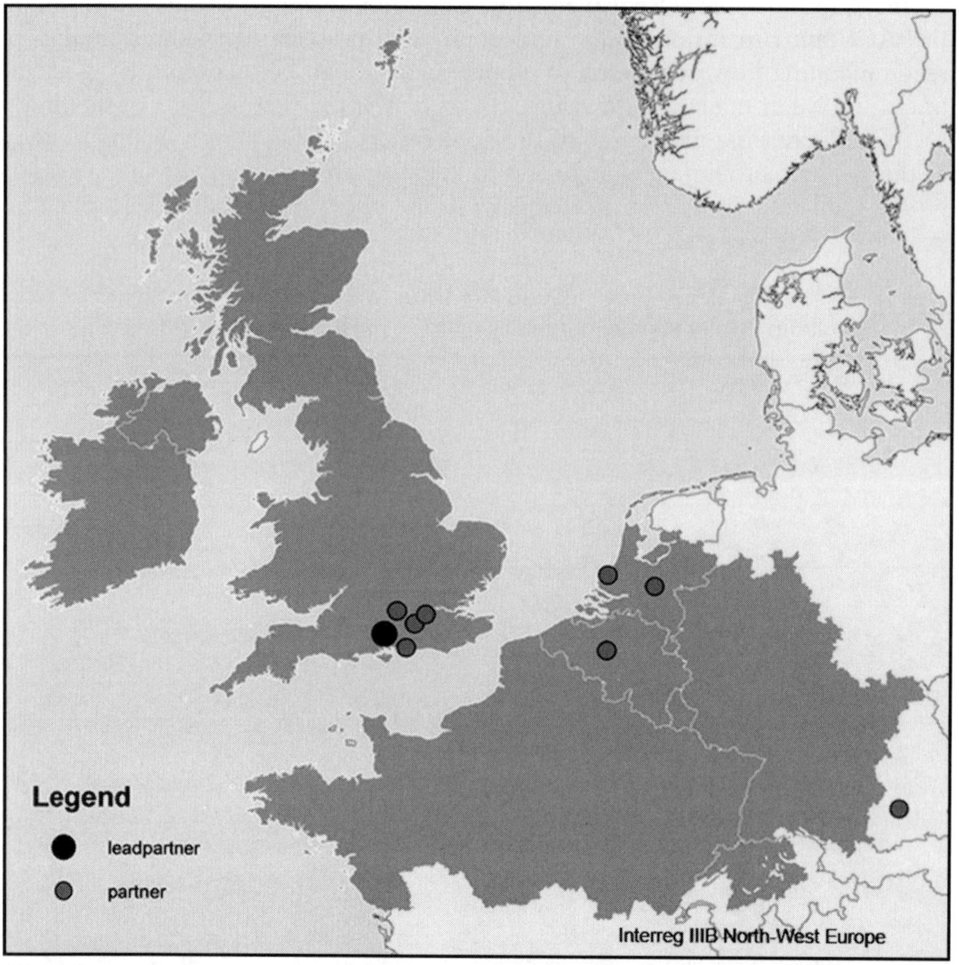


ESPACE has also established an 'Extended Partnership' – a targeted network of stakeholders who are interested in being involved in the development of the outputs of ESPACE and who can act as ambassadors for the project. There are now over 200 members, including representatives from the European Commission, European Environment Agency, national, regional and local governments, non-government organisations, universities, research institutes, consultancies and other EU projects.

A range of actions have been undertaken by partners to identify how to best adapt to climate change. These include actions looking at the most effective ways of raising awareness, the role of behaviour change in adapting to climate change, policy review through the use of innovative techniques and the development of tools and models.

## Behaviour change work paves the way for effective action on climate change

It is very easy for organisations and individuals to fall into the trap of believing that 'raising awareness' of climate change issues will somehow automatically be a precursor for the changes in lifestyle choice and behaviour that need to occur. Conversely, evidence shows that this is not the case. In reality, raising awareness of climate change issues without paying careful attention to other factors can be overwhelming for many people, and will often reinforce people's defences not to deal with the problem. This is because awareness raising on its own can lead to difficult emotions and unhelpful responses. Developing realistic, long-lasting, and effective actions on climate change requires a more sophisticated understanding of the role that 'awareness' plays. To tackle these issues, Hampshire County Council (HCC) has undertaken research to understand more about how to bring about action and the role that local authorities can play in doing so.

Alexander Ballard Ltd, experts in the field of behavioural change and climate change, were commissioned early on in the project to help ESPACE address this important issue (ESPACE 2005a). Their expertise has provided the project with a greater understanding and has informed the way in which climate change is now communicated.

One of the most important lessons ESPACE has learned from this work is that there has to be a balance of three conditions before effective action is likely, and that a conscious effort to maintain this balance will help campaigns to succeed. One of these three conditions is 'Agency' – the sense that what an individual is doing is worthwhile and will make a difference in the face of a potentially overwhelming issue. Another condition is 'Association', by which we mean participation in groups. Association is needed to develop 'Agency' as well as reinforce changes in behaviour. The third condition is 'Awareness', which, if developed when requested in a timely and accurate fashion, will lead to appropriate responses being developed.

There is one final condition that is required in order to create a truly robust cycle of action-focused responses – the process of 'Action & Reflection'. This is a cyclical process of reflecting upon actions in order to maintain meaningful responses. This will help to ensure that bad habits are not repeated and that good practice is shared and developed. The 'Action & Reflection' process combined with the three conditions mentioned above have been collectively dubbed the '4As' (**A**gency, **A**ssociation, **A**wareness, and **A**ction & Reflection). The appreciation of how these work together has improved our understanding of the role that local authorities can play in encouraging appropriate action orientated responses.

Fig. 3: ESPACE partners applying the behaviour change principles to a live case study. Credit – ESPACE 2006

Armed with this knowledge, HCC began to ask more questions about other change factors. This then led to another commission with Alexander Ballard Ltd, looking into the role of 'champions' in bringing about and supporting change (ESPACE 2005c). By 'champions' we mean people who have deep awareness of climate change issues and persistently look for 'Agency'.

From this work we discovered that there are a number of identifiable types of champions and that providing support for champions on climate change issues is vital for success. It is evident that champions will often take a wider approach to issues that limit change and can therefore often develop innovative ways to deal with existing constraints. What became clear is that where there is no champion, there is no change in behaviour. The concept of developing and supporting an 'ecosystem' of champions is now being used as an approach by HCC in the development of its own Climate Change Action Plan.

In addition to this work, HCC is also working with Alexander Ballard Ltd and ESPACE Partner, South East Climate Change Partnership, in developing a tool that will allow organisations to visualise and assess their ability to adapt or 'adaptive capacity'. The tool will take the form of a matrix that will allow organisations to benchmark their capability to respond to climate change, as well as encourage good practice by demonstrating what exemplar behaviour might look like. Steps on how to improve performance will be contained within the matrix, which in turn will allow organisations to visualise how different actions and pathways work together and where they should prioritise actions and resources for dealing with climate change. This work is currently ongoing and will be completed at the end of the project extension in the summer of 2008. The behaviour change research has been very well received by many organisations across Europe and it is expected that its application will have far reaching implications for assessing the performance of organisations.

## Raising awareness – developing effective ways of communicating climate change

Climate change is too abstract for the general public to relate to. An average rise of temperature of a few degrees spread over one century isn't visible for the people in the street and makes it difficult for them to understand the consequences or to find a solution. The media have only compounded this problem by focussing on 'doom and gloom' headlines.

Regionaal Landschap Zenne, Zuun & Zoniën (RLZZZ) recognises that this 'climate fever' has a detrimental effect on the public, leading to a 'head in the sand' approach. It is clear that our approach needs to change and RLZZZ believes that there are huge gains to be made in educating and influencing the younger generation. They are the decision makers of the future and will be faced with adapting to the climate change impacts we have initiated.

For pupils aged 16–18 in Belgium, spatial planning and climate change issues are both part of the official school programme, although they are not yet connected. RLZZZ have developed 'Caring for Climate' – a climate championship for schools to bring the two issues together, helping the younger generation understand climate change and how to respond to the impacts.

Inspired by the '4 As' principle developed in ESPACE – Awareness, Agency, Association and Action & Reflection, (ESPACE 2005a), the 'Caring for Climate' championship aims to turn negative thinking on climate change into positive action. Schools can participate by entering a 'Climate Team'. Coached by a geography teacher or 'Climate Coach', each team is tasked with developing an educational package called a 'Climate Kit'. The kit can be a game, DVD, website, song, publication or exhibition. The winner of the championship is awarded the title of 'Climate Champion' at a final event that includes a ClimXpo and a ClimaXparty. The ClimXpo is an exhibition showcasing the Climate Kits developed by each school, eco-minded companies and spectacular weather photography. The exhibition finishes with a ClimaXparty that includes a number of fun, low carbon games for the children, for example: 'Dyno 2' – a bicycle which can drive a music installation, a 'Juice Bicycle' that creates biological fruit juice and a 'Sun Mobile' that delivers warm water to the dishes.

After the ClimXpo event, the Climate Teams are taken on range of educational trips to help stimulate behaviour change. The winning team will be taken on a 'Sunny Route' – a trip starting from the Ecoboat Isera, where they learn about energy efficient and renewable technology. Using electric bikes the teams then visit several ecological projects including a water purification system, an authentic windmill and a low energy house. A trip is also made to the Royal Meteorological Institute of Belgium (KMI), giving the pupils an understanding of forecasting techniques and the science of climate change.

The concept of the championship is based around an original principle developed through the ESPACE behaviour change work – 'taking action as a starting point, then attaching awareness on to it, is far more effective than developing awareness in isolation' (ESPACE 2005a). Using this important rule in the Climate Championship has helped to stimulate competition, initiative, dialogue, co-operation and behaviour change amongst pupils and will ultimately improve their capacity to deal with the impacts of climate change.

In addition to the awareness-raising work carried out by RLZZZ, West Sussex County Council (WSCC), have been running an awareness-raising campaign called 'A Climate for Change'. There are two aspects of this campaign; the first involves assessing changes in the level of understanding of climate change across West Sussex in response to awareness raising campaigns and the second involves a case-study for the Manhood Peninsula – a low lying, vulnerable coastal area in West Sussex.

Following an initial baseline survey undertaken in 2004, which revealed a low level of awareness of climate change in West Sussex (ESPACE 2005b), the project has aimed to identify, develop and implement the most effective methods of increasing public awareness. This survey has recently been repeated and has shown that understanding about climate change in West Sussex has now improved. The work also suggests that the most successful methods have been those that fully engage individuals (ESPACE 2006f).

Across West Sussex, schools have always been a key target audience for awareness-raising initiatives. During the project, over 75% of secondary schools were represented at a schools' climate change conference at which students were able to meet and learn from the experiences of young people from all over the county. The conference was so well received, that an 'Eco-fund' has now been established to enable school councils to carry out climate change projects. Another school-based project, 'Formula Sun', which involves pupils in designing, building and racing solar power cars, has also demonstrated the importance of engaging children in an interesting and fun way in order to raise awareness of the impacts of climate change.

Fig. 4: The Manhood Peninsula, Selsey, West Sussex, is one of the lowest-lying coastal areas in the UK and is the focus of a case-study looking at effective ways of engaging with the local community. Credit – West Sussex County Council/PPL

Engagement has also been the key to other successful projects. The Eco-faith school project for example has engaged faith groups in the subject, providing an opportunity to show how climate change is relevant to their work. The '10% Challenge' project has engaged the wider population of West Sussex on climate change by challenging people to reduce their environmental impact by at least 10%. The 10% Challenge also highlighted the importance of political support for such initiatives, which has been critical to this project's success.

The case study on the Manhood Peninsula illustrates how education and awareness-raising can be delivered through active engagement of people in planning for climate change. It has been developed through the creation of a local 'Adaptation Action Plan', containing a set of social, environmental and economic actions, that together form an integrated framework for adaptation (ESPACE 2006a).

Whilst the Action Plan will help the Manhood Peninsula to prepare for climate change, it was the *process* of developing the plan that served as a valuable educational tool. This process was based around a series of public workshops and the use of an open dialogue approach that allowed for a two-way exchange of information, knowledge and ideas between policy and decision makers, businesses and communities. To engender a strong sense of ownership of the project outcomes, this open dialogue was initiated right from the start. The workshops provided the opportunity for direct communication between those who decide and those who are affected by planning decisions. As a result of the co-operative and open nature of the process, a strong and widely-accepted common agenda emerged.

The launch of the project included a Strengths, Weaknesses, Opportunities and Threats (SWOT) analysis of the local area which provided the baseline information upon which the project would develop. To create a common level of awareness amongst participants, the public workshops were preceded by the dissemination of topic-specific background papers, developed from the SWOT information. These papers were visually presented at the beginning of each workshop and participants were given the chance to use their own knowledge and experience to alter the content. This process was inclusive and ensured that the issues considered were of local relevance and therefore 'real' to the participants. It also re-inforced ownership of the content amongst participants.

A number of interactive and visual techniques were used in the workshops, including 'future-basing' – a technique that proved to be an excellent way to picture what climate change could really mean for the local area. It generates a picture of a successful future, sparks creative thinking and seeks to generate a sense of common purpose between the people involved. 'Options for action' sessions then focused on identifying various approaches to achieving a vision, exploring the pros and cons of particular approaches in terms of long-term resilience, social, economic and environmental impact.

A final 'integration workshop' was designed to raise awareness of conflicts between various adaptation options and to build an integrated response to adaptation. Four scenarios were derived from two locally relevant climate variables distinct to the area. Participants were given time to use large maps of the area to physically explore the scenarios and identify the conflicts between different land-use claims. This gave participants the skills and understanding to make informed decisions on responses appropriate to the local area.

To retain the dialogue and the focus on appropriate local responses, workshop outputs were then refined by smaller action planning groups with both local and technical representation. Those participating in the sessions were given the brief to turn the workshop outputs (the 'what') into action programmes (the 'how'). Following responses from a public consultation phase, the refined action plan was adopted by the local steering group, the Manhood Peninsula Partnership.

The content of the action plan emerged from the input of both professionals and residents, and this collaboration has been central to the project. Structures for managing and delivering the action programmes have been designed to ensure that the dialogue developed through the engagement process is sustained, and that climate change adaptation on the peninsula continues to be relevant, accepted and sustainable. The action plan is considered to be a 'working' document, and a rigorous review and updating process has been incorporated into the delivery phase.

## New partnership approaches for engaging stakeholders

Another method for delivering key messages, raising awareness and influencing behaviour change of individuals is the effective use of climate change partnerships. South East Climate Change Partnership (SECCP) is developing a partnership model that can be used to raise awareness and ensure that there is more acceptance to adapting to climate change amongst planning practitioners and policy makers. SECCP itself is an example of a successful regional climate change partnership that has engaged with a range of stakeholders from the public, private and voluntary sectors and has also raised awareness, generated new projects and research.

Within ESPACE, SECCP is developing a transferable model for a regional climate change partnership that other European regions can adapt to their own circumstances. Partnership working is particularly suitable to adapting to unavoidable climate change, not least because:

- it has direct and indirect impacts on a very wide range of organisations, communities and geographical areas
- the information that exists on climate change is often very technical and needs experts to interpret;
- the issues on which climate change impacts are also complex and subject to many other pressures;
- no organisation has complete control over those factors that could reduce its vulnerability to climate change;
- climate change may present opportunities for some organisations or sectors while presenting risks to others – and the responses of one will interact with others.

For many organisations and individuals, climate change is often seen as a long-term issue rather than an immediate priority. This is likely to lead to a reactive rather than a proactive approach. Partnership working should help to change these perceptions, stimulating a more integrated approach over time.

Building on the outputs of ESPACE as well as SECCP's own development over five years and the experience of other regional climate change partnerships in the UK, the partnership has recently adopted a new three year work programme and communications strategy. Key principles of the programme and strategy include a need to recognise that:

- *We need to relate climate change to people's existing or emerging priorities* – Climate change impacts in different ways on the immediate priorities of a wide range of people – but is not necessarily the key issue for them.
- *We need to address the context for change as well as the priorities* – People have critical barriers and springboards which affect how they engage with change. This will set a context for finding and delivering meaningful responses to climate change 'problems'.
- *We need to identify and develop opportunities for real agency for change* – Raising awareness is necessary but is not sufficient on its own to deliver mitigation or adaptation. Increasing awareness can leave people with the overwhelming feeling that the problem is too large for their own actions to have any effect, or that it is too confusing, is already decided or is inherently intractable.
- *We need to identify and support existing champions and develop new ones* – Change needs champions. Commitment and strategic skills are essential to translate climate change awareness into concerted action within a given organisation or environment.
- *In order to demonstrate and promote good practice, we need to reinforce positive association between people and help them to demonstrate the value of the changes they are making* – Improved practice spreads very slowly where the benefits aren't clear, perhaps as a result of poor levels of association or mutual support between practitioners.
- *At the partnership level, our action is to learn and help others to learn* – Although dialogue and association between people and organisations are invaluable, higher levels of awareness usually result from action, rather than the other way round.
- *We need to focus on helping people to build their capacity, so that they can deliver adaptations that are appropriate to them* – Building 'adaptive capacity' is different to delivering adaptation. While some people and organisations already have the capacity to deliver some actions, everyone can benefit from doing more to develop their capacity further.
- *We need to focus on the value that partnership working can add to partners' individual actions* – Partnership resources are often limited – in terms of money, staff and the capacity to engage in joint working. Partnerships should aim to harness and add to what their partners already do or can do – not try to replicate or substitute for them.
- *We need to capture and 'brand' people's successes in climate change adaptation and mitigation – and celebrate these.* This will help those delivering ac-

tions to connect their successes to the bigger picture. Small successes count in their own right because they help us move forward and generate larger ones.

- *We need to share and develop a programme of work that helps individuals and organisations to improve their actions over time.* Responding to the challenges of climate change is a process of development. We cannot expect to identify all the necessary actions now – the process of engaging in change leads to improved actions.

By developing cooperation and awareness among key stakeholders (ESPACE 2005a), partnerships can provide an essential part of the architecture for change that we need to build if we are to successfully adapt to the new climate that is now inevitable over the next 50 years and beyond.

## Development of Tools and Models for spatial planners

Co-inciding with the awareness raising and behaviour change work, ESPACE has developed a suite of new tools and models for policy makers and practitioners to test existing policies and to look at how they stand up to the impacts of climate change. The models will be designed to test common practices in dealing with problems such as water resource issues, flooding and land use planning in a changing climate and will also examine historical evidence and experiences.

A key component of this work, lead by the Environment Agency (EA), is the development of a Decision Testing Framework that enables planning decisions to be simulated and tested against a range of future climate change scenarios. Building on the framework produced by the UK Climate Impacts Programme and various tools already used by EA such as the MDSF (Modelling and Decision Support Framework) , the new framework allows the costs and benefits of strategic flood risk options to be tested (ESPACE 2006d). These tools developed in the UK have been compared to the approaches to climate change modelling developed by other ESPACE partners in Germany and the Netherlands to ensure the transferability of the various models across Europe (ESPACE 2006d). Building on this work, the EA are now developing a decision testing tree to allow decision-makers to follow through the consequences of a series of decisions taken over time in the face of changing climate situations.

Fig. 5: This image from the Decision Testing tool illustrates overtopping and flood depths from an event in 2050. Credit- Environment Agency 2006. – Future Scenario 2050 – Deterministic Overtopping Flood Extent (1:10,000yr Event)

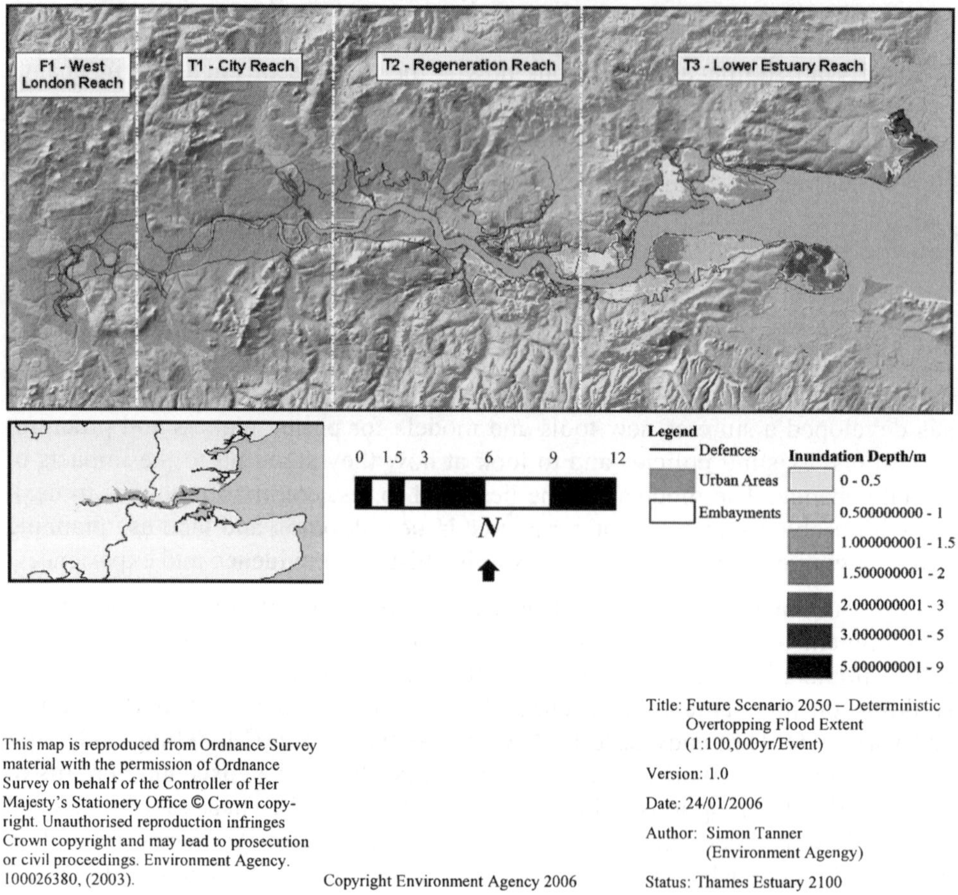


The Bayerisches Landesamt für Umwelt (LfU) is using the flood-prone catchment of the Fränkische Saale in Germany as a case study area to develop spatial planning tools and solutions for adapting to climate change. With the threat of an increase in extreme weather events and flooding as a result of climate change, Bad Kissingen's Water Management Agency are examining various planning and design options to minimise flood damage now and in the future e.g. the creation of retention spaces. Climate change is being taken into account in the decision making process using the tools developed by LfU in ESPACE. To support this work, LfU have creating a suite of models that simulate the impacts of regional climate changes on hydrological parameters, e.g. water levels and flood frequencies.

LfU have recently launched a website for the Fränkische Saale (www.klimaprojekt-espace.bayern.de). Developed to raise awareness of individuals and local authorities, this site includes information on climate change impacts and flood protection measures as well as a series of flood risk maps that take climate change into account. This will help individuals improve their assessment of flood risk and in turn develop appropriate adaptation responses. The website also contains information from the first stakeholder meeting that engaged local decision-makers in the project. The event held in February 2006 sparked off strong public interest in the project and resulted in a flurry of newspaper articles and a short television documentary.

Fig. 6: Flood maps have been created using climate change data from ESPACE models. Credit – LfU

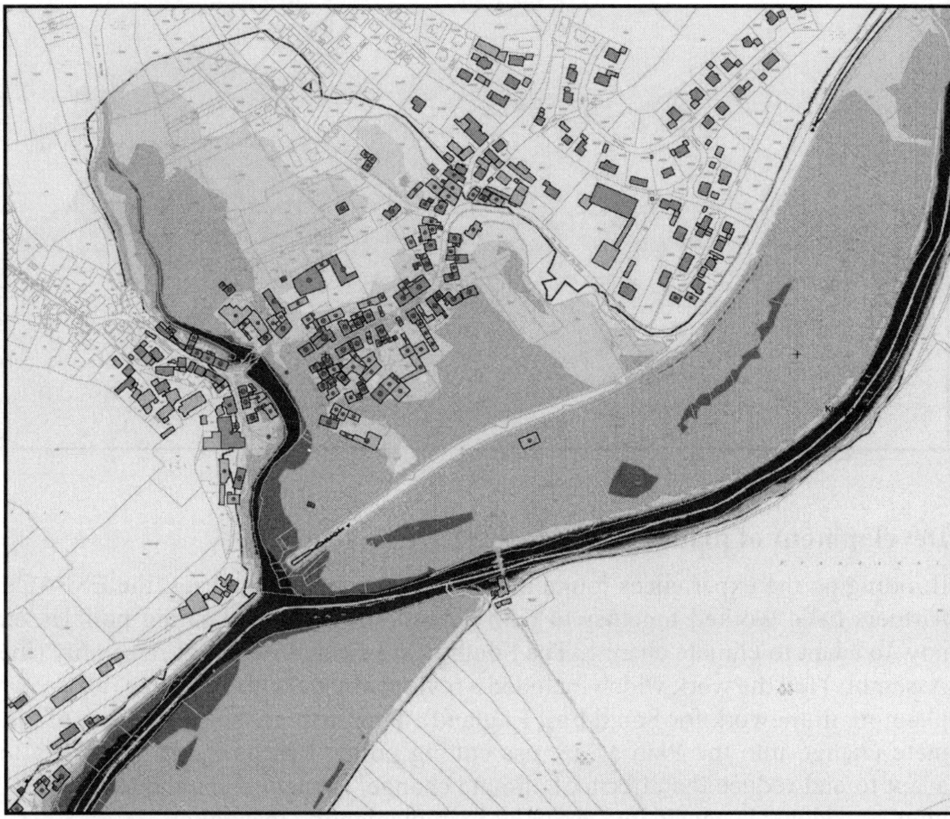

Waterschap Rivierenland, have also developed a range of models, focussing more on the predicted impacts of climate change on water resources. These 'Guiding Models' can be used in regional planning in the search for water storage areas (ESPACE Project 2004). In these models, several functions are interconnected to water such as ecology, cultural history, agriculture, tourism and urban issues. This ensures that the models can be used as a holistic tool for planning by taking the characteristics of the local environment into account in the decision making process. A handbook has now been produced that will allow the user to choose which Guiding Model is most relevant to their geographical location and planning system.

Fig. 7: An example of one of the components of the ESPACE Guiding Models. This is used to build up the characteristics of an area. Here you can visualise the interplay between land use functions. The blue surfaces are the lowest parts which are most suitable for water storage. Credit – Robbert de Koning, Oosterbeek, The Netherlands.

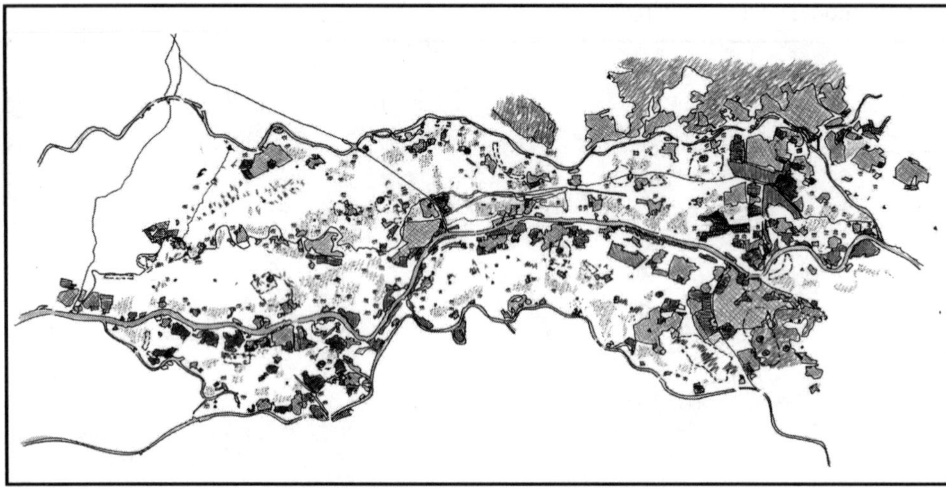

## Development of planning policies

Building on the experiences found in each of the partner countries, the ESPACE partners have worked together to help determine regional planning policies on how to adapt to climate change. The South East England Regional Assembly (the Assembly) led the work which included a review of the South East Plan (a 20-year planning framework for South East England). This ensured the integration of climate change into the Plan as a cross cutting policy that promotes measures to adapt to and reduce the effects of climate change. Behavioural change of the region's residents is considered essential in implementing this policy.

Following on from this, a dedicated 'Climate Change Mitigation and Adaptation Implementation Plan' has been developed (ESPACE 2006c), providing guidance and actions on how to deliver the adaptation policies within the South East Plan. Proposed activities cover a variety of measures including:

1. Guiding development to areas where climate change has less impact, i.e. areas protected from flooding, erosion, storms, water shortages and sinking of the land level;
2. Ensuring buildings are resilient;
3. Incorporating sustainable drainage and water efficiency systems into developments – for example reed beds or low-flush toilets;
4. Increasing flood storage capacity and developing new water resources;
5. Ensuring options for sustainable flood management;
6. Considering impacts on wildlife habitats and species.

For each of the above, the Climate Change Implementation Plan identified key issues and reviewed the effectiveness of existing actions such as the available research, raising-awareness issues, partnership working, and ways of developing and implementing policy. This provides a useful audit of actions, which is currently being undertaken by the Assembly. The Implementation Plan also identifies the barriers to implementation including:

1. Uncertainty about the nature, complexity and extent of climate change and its impacts
2. Regulations, codes and standards which have not yet been revised to reflect anticipated climate conditions
3. Conflicting priorities, perceptions, behaviour and attitudes and the lack of incentives to change these

This information is used to prioritise actions per sector at a national, regional and local level and to identify the relevant organisations that are responsible for these actions.

The Assembly has also prepared a good practice guide 'Adapting to Climate Change Impacts on Water Management: for Planners and Developers' (ESPACE 2007a). Planners need to use the guide when developing spatial plans. This way developers will be required to undertake water management adaptation measures described in the guide. Matters such as pressure on water resources, flood risk and resilience to other water-related climate change impacts are addressed.

Case studies with useful tips from a number of locations in the South East of England and different types of development are covered in the guide. It also includes seven principles to help planners and developers integrate adaptation to climate change into planning documents:

1. *From the start* – Consider measures for climate change adaptation throughout the planning process. This will avoid adaptation becoming an 'add-on'. It helps to minimise the disruption and delay of the process and will ensure that the process is more effective.

2. *Be flexible* – Flexibility of design is crucial. Developments should be resilient to a range of future climate scenarios. Adaptation options might offer synergies (supporting wider environmental or adaptation outcomes) or conflicts (working against other possible adaptations).
3. *Maintenance* – Consider the need for on-going maintenance and appropriate end-use. There may be a need to explain how to maintain special systems to the occupiers of developments e.g. Sustainable Urban Drainage Systems must be de-silted and monitored.
4. *Integrated approach* – Integrate climate change adaptation and mitigation measures wherever possible. There are some instances where adaptation options can help reduce greenhouse gas emissions, e.g. green roofs.
5. *Be innovative* – Innovation can lead to wider benefits. Case studies demonstrate the benefits of successfully integrated adaptation measures in existing development.
6. *Use all available information* – Existing tools and regulation can help support the adaptation response. You can address climate change adaptation through a range of mechanisms, including Sustainable Environmental Assessment, Building Regulations and Planning Consents.
7. *Partnerships* – Partnership working adds value. Working in partnership with organisations and groups such as the Environment Agency, the Government Offices, regional planning bodies and regional climate change partnerships can support local councils and provide advice with integrating adaptation into the planning and development process.

The Implementation Plan and the Guide for Planners are not bound to the UK and can be used elsewhere in Europe. They will help practitioners and policy makers identify ways to help reduce the impact of climate change.

Building on this regional policy development, Ministerie van VROM organised two International Conferences in 2007 that focused on developing robust adaptation policies at the national level. The first of these events took place in early spring 2007 with the aim of:

- exchanging knowledge and experiences on national adaptation policies to climate change
- comparing national adaptation strategies
- testing the draft Dutch national spatial strategy on adaptation to climate change
- preparing the international stakeholder congress (May 2007) on adaptation in the Netherlands.

The results of the conference were used for refining the Dutch National Adaptation Strategy and have contributed to the development of the final ESPACE

Common Transnational Strategy for adapting to climate change. The second event was an international congress that aimed to:

- enhance the level of awareness of international stakeholders in respect to adaptation to climate change;
- exchange knowledge and experience on adaptation;
- deliver international commitment for adaptation on climate change;
- present the Dutch national adaptation strategy on climate change.

An analysis of all of the above Partner actions was undertaken by "acclimatise Ltd" in 2006 to help draw out the key transnational findings and lessons learned throughout the duration of the project (ESPACE 2006e). This review focused on:

- How to manage climate adaptation through spatial planning
- Developing and applying adaptation policies and strategies in spatial planning
- Methods so that the planning process can foster and promote adaptation
- Creating capacity and understanding with key stakeholders of how spatial planning mechanisms can be influenced to address climate adaptation
- Awareness of climate adaptation issues and creating 'agency for change'.

These lessons have been used to develop a common transnational strategy that outlines the ESPACE project recommendations for incorporating climate adaptation into spatial planning (ESPACE 2007b). The strategy includes a number of key recommendations for individuals and organisations involved in spatial planning processes, for example:

- Adopt two main approaches to deliver a climate-adaptive spatial planning system:
  - approaches based on *change management* – to create the governance and management required to address adaptation in spatial planning;
  - *Risk management* processes – to integrate climate risks into the process of policy – making and identify the most appropriate adaptation policies and measures;
- Commit to taking account of changing climate risks over the whole of the decision 'lifetime' (e.g. the lifetime of a built development) rather than over the spatial plan period;
- Create a conscious balance of awareness, agency and association, to encourage, inform and reinforce long-lasting actions on climate change adaptation. Foster 'climate adaptation champions' who have a key role in creating and sustaining momentum.

These recommendations are then backed up by guidance, policies and tools developed in the ESPACE Project that illustrate how to deliver the key messages.

Through ESPACE, the Partnership has become aware of obstacles to the delivery of adaptation to climate change, even where appropriate policies are in place. A twelve month extension to this project will now explore these barriers and develop new methods for successful implementation of adaptation policies.

All of the actions undertaken during this project will provide the foundation from which spatial planning policy recommendations that take account of climate change can be developed. It is hoped that this new transnational approach will result in a change in policy practice for the whole of Europe by ensuring that climate change is acknowledged as a major influence on spatial planning decisions and processes.

## References

*Bhattacharya, S.* (2003), European heatwave caused 35,000 deaths, New Scientist – 10/10/03.

ESPACE Project (2004), Guiding models for water storage. (In English and Dutch). www.espace-project.org.

– (2005a), How can local authorities stimulate & support behavioural change in response to climate change? A report for Hampshire County Council undertaken as part of the ESPACE Project (vol. 1 & 2). www.espace-project.org.

– (2005b), Measuring Awareness of Climate Change. Report on Stage 1 of ESPACE project Adapting to Climate Change: Raising Community Awareness in West Sussex www.climateforchange.org.uk/ws%20results.asp.

– (2005c), Warm Hearts and Cool Heads: The Leadership Potential for Climate Change Champions. A report for Hampshire County Council (vol. 1 & 2) www.espace-project.org.

– (2006a), A Climate for Change on the Manhood Peninsula – Manhood Peninsula Adaptation Action Plan www.climateforchange.org.uk/manhood_action_plan.asp.

– (2006b), A Toolkit for Delivering Water Management Climate Change Adaptation Through the Planning System www.espace-project.org.

– (2006c), Climate Change Mitigation and Adaptation Implementation Plan – www.southeast-ra.gov.uk. and www.espace-project.org.

– (2006d), ESPACE Decision Making Framework and Tools – Phase 3 International Comparison. Final Report. www.espace-project.org.

– (2006e), ESPACE Project: Analysis of Partner actions. Final Report. October 2006. www.espace-project.org.

– (2006f), Measuring and Tracking Climate Change Awareness in West Sussex. Report on Stage 2. Adapting to Climate Change: Raising Community Awareness in West Sussex. www.climateforchange.org.uk/ws%20results.asp.

– (2007a), Adapting to Climate Change Impacts on Water Management: A Guide for Planners. www.southeast-ra.gov.uk and www.espace-project.org.

– (2007b), ESPACE Project: Final Common Transnational Strategy (in preparation at time of writing).

Met Office UK (2005), Annual report – 2004/5.

*Stern, N.* (2006), The Economics of Climate Change – The Stern Review, Cambridge University Cabinet Office – HM Treasury, Cambridge University Press www.hm-treasury.gov.uk/ independent_reviews/ stern_review_economics_climate_change/sternreview_index.cfm.

*Stott, P.A., Stone, D.A. & Allen, M.R.* (2004), „Human Contribution to the European Heatwave of 2003," Nature, vol. 432 (2 December 2004), pp. 610–13.

Chapter 2

# Communicating Climate Change Impacts – An Approach Derived from two Regional Development Projects

**Philipp Schmidt-Thomé[1], Johannes Klein[1]**

## Abstract

This chapter analyses the results of climate change communication between researchers and practitioners of two EU-funded regional development projects („Sea level change affecting the Baltic Sea Region“ SEAREG & „Developing Policies & Adaptation Strategies to Climate Change in the Baltic Sea Region“ ASTRA). The interesting observations made by both projects have led to a deep insight of the practical constraints and opportunities, as well as the risks, in communicating climate change between different stakeholders. An analysis shows „wake up calls“ are often required to raise interest on potential climate change impacts. Further research therefore should focus on streamlining climate change communication. This chapter develops a theoretical background, which could find practical approach in future projects.

## 1. Introduction

This chapter analyses the results of climate change communication between researchers and practitioners of two EU funded regional development projects („Sea level change affecting the Baltic Sea Region“ SEAREG & „Developing Policies & Adaptation Strategies to Climate Change in the Baltic Sea Region“ ASTRA). Both projects focused on the Baltic Sea Region, and meanwhile the first project focussed on the effects of climate change on sea levels, the second one aimed at a broader scope of climate change impacts. The interesting observations made by both projects have led to a deep insight of the practical constraints and opportunities, as well as the risks, in communicating climate change between different stakeholders. It is certainly true that the motivation to develop adaptation strategies in climate change is based on different backgrounds on all involved actors. But it was also revealed that, even though the discussion on climate change is dominating the news, at least since the Kyoto protocol and the Hurricane Katrina, the knowledge upon potential climate change impacts on a local scale are mostly perceived on recent extreme events and resulting public pressures, rather than on an integrated analysis.

1 Geological Survey of Finland.

## 2. Lessons learnt from SEAREG and ASTRA

The „Sea level change affecting the Baltic Sea Region“ (SEAREG) project was co-financed by the European Regional Development Fund (ERDF) INTERREG III programme and focused on sea level changes in the 21st century (Schmidt-Thomé & Peltonen 2006, www.gtk.fi/slr). It used climate change scenarios developed by the Intergovernmental Panel on Climate Change (IPCC), regionally downscaled, to display sea level changes in local case study areas. The project developed three sea level change scenarios, for the entire Baltic Sea Region and for each case study, named „low“, „ensemble average“ and „high“ (Meier et al. 2006). The intention of using three scenarios was not to propose that the ensemble average scenario would be the most appropriate, but to display the range of uncertainties, according to present knowledge. Three scenarios graphically display what the range of changes might be and what not, i.e. they underline that the sea level will rise (and is already rising) in any case, but that there is also an upper limit, at least according to present scientific knowledge (Schmidt-Thomé 2006). The extent of the sea level changes might be different if analysed under the recently published summary for policymakers of 4th assessment report of the IPCC (2007), but the key message stays: there will be changes in sea level and stakeholders should get prepared for it.

In addition to the changes in sea level, the respective changes in flood prone areas were displayed on the case study maps of the SEAREG project. The assumed flood levels were an average of recorded flood levels, as extreme cases were not used to avoid too drastic scenarios. Nevertheless, the project was often strongly criticized by both scientists and planners, that especially the high case scenario would rather be a „horror scenario“ as it was impossible for the sea level to rise to such extents.

The criticism petered out after the winter storm „Gudrun/Erwin“ swept through the Baltic Sea Region in January 2005. One of the case study areas of the SEAREG project was the Estonian town of Pärnu. It is a low lying town, situated on quaternary deposits on the west coast of the Pärnu bay. The slowly flowing, meandering Pärnu river cuts the city in two parts before discharging into the Baltic Sea (see figure 1 below). The town is strongly dependent on tourism sector and its long, white sandy beaches are important for the local economy (Klein & Staudt 2006, www.gtk.fi/slr). The 2005 winter storm lead to record flood levels in Pärnu. The flood levels were as high as those displayed for the high case scenario for a 100-year sea level change and the additional flood prone area . Although the SEAREG project emphasised that this storm surge was an extreme event that was not necessarily triggered by climate change, the storm dramatically showed that the flood levels that are currently possible in extreme cases, might well be repeated in the future.

Pärnu is a case study also in the SEAREG follow-up project, „Developing Policies & Adaptation Strategies to Climate Change in the Baltic Sea Region“ (ASTRA, www.astra-project.org). It was due to the scenarios developed in the SEAREG project and the further participation in the ASTRA project, that a decision taken by the Pärnu Town Council on starting to construct flood defence systems along the shoreline and the riverbanks was halted by the Town mayor. The decision of the Town mayor reads that before any flood defence structures would be erected, the research results of potential effects caused by a changing climate should be taken into account, precisely naming the ASTRA project (Pärnu Town Government 2006). This indicates that the Town Government plans to undertake long-term adaptation measures instead of short-sighted reactions.

But what has led to the reaction of the Town Government? And how long-lived will this new action plan be? According to information given to the ASTRA project „off the record“ the decision of delaying flood protection measurements might have been taken because of other, rather short-termed political interests. The fact is that a flood protection for Pärnu Town will be both very costly and very difficult from an engineering point of view.

The original layout of the ASTRA project was to assess climate change impacts that go beyond sea level rise and coastal flooding, for example, storms, droughts, landlsides, etc. The ASTRA project also thought of following a different approach on communicating climate change effects to practitioners. The SEAREG project intended to find appropriate ways of displaying and jointly analysing expected impacts, but encountered great challenges because first of all the concepts of climate modelling and uncertainties had to be clarified (K. Schmidt-Thomé & Peltonen 2006, Lehtonen & Peltonen 2006). Also, as mentioned above, even modest sea level rise and flood prone area change scenarios faced a strong rejection, probably simply because people could not imagine the potential impacts. Therefore, the public recognition of the SEAREG project clearly benefited of the 2005 winter storm. The assumption for the ASTRA project was therefore that, once the winter storm had occurred, the addressed stakeholders would be more informed about climate change and its potentials. The project's approach was thus to ask the practitioners for their point of view on potential climate change impacts and the possible effects for their respective case study areas. A questionnaire was prepared at the beginning of the project to detect these and derive the further project activities on communicating and displaying the potential climate change impacts accordingly (Eisenack & Kropp 2006).

But this strategy also resulted in a challenge, because it mostly focused on climate change impacts of sea level rise and coastal flooding, while all other potential impacts were simply left out. It was therefore necessary to adapt the new strategy and return to one of the SEAREG project's aims namely with the handling of basic information. And here again certain disbelief was encountered, simply because there were no references too catastrophic events on which assumptions could be practically based on. The project, which started in 2005, suggested that even though most climate change scenarios propose an increase in precipitation for northern Europe, there is also a potential for an increase of extreme events (Bärring & Persson 2006).

In the case of Finland one proposal was to study the effects of droughts on regional development. Even though Finland is often referred to as the „country of the thousand lakes“ (Finnish Tourist Board 2007), the country has experienced some droughts. The average depths of the lakes are shallow and droughts can have profound impacts on water availability and quality (Kuusisto (ed.) 2006). Nevertheless, drought was not perceived as a major problem by stakeholders involved in the ASTRA project. This until the summer of 2006, when Finland received less than 50% of its average precipitation (Ilmatieteenlaitos 2006). This has led to strong shortages of water for the agriculture, and especially in regions with shallow aquifers restrictions of domestic water use were issued.

The above mentioned climate change effects on precipitation and extreme events mean that, in the case of droughts, it could also be possible to have a couple of extremely dry summers, even though the average precipitation might increase over the next century. Thus, the effects of droughts should be taken into account by preventive measures, e.g. storage of sufficient water in times of water abundance. These decisions must be taken on a binding level in order to ensure water availability and the ASTRA project is continuing to press these issues forward in its respective case study areas.

The lessons learned from the SEAREG and ASTRA projects so far is that both approaches have only been partly successful, as it required extreme events that issued „wake up calls“ concerning potential climate change impacts. Further research therefore focuses on streamlining climate change communication. The following section intends to develop a theoretical background, which could find practical approach in future projects.

## 3. Outline of a conceptualised climate change communication

### a) Geographic context

Lehtonen and Peltonen (2006) ask for a contextualisation of the scientific knowledge that goes beyond the topography and takes into account local needs and values. A (very small and very incomplete) review may shape a picture how the contextualisation can look like. Fleischhauer (2004) identified three major criteria for natural hazards, their perception and relevance for spatial planning. The impact has to be bound to location, the impact can be delineated and the impact can be influenced. All three criteria can be applied as well to many impacts of climate change mentioned above. They take place in a geographic context. Therefore, the impacts of climate change can be taken into account on the local level and adaptation strategies can be developed, i.e. the impact can be influenced. However, in the context of ASTRA these criteria (bound to location, impact delineation and influence on the impact) might be too narrow, since they are valid mainly in relation to spatial planning.

The feedback to questionnaires of the ASTRA project (Eisenack & Kropp 2006) shows that climate change is discussed on the local level often more in the context of mitigation than adaption (following the distinction of mitigation and adaptation of Smit et al. 1999). When it comes to adaptation, typically climate change impacts related SLR, storm surges (and erosion) and changing probability for floods depending on a change of precipitation patterns are mentioned. All these effects (SLR, storm surges, erosion or floods) have a well-defined geographic dimension and can be delineated on a map.

### b) Social and socio-economic context

The relation to the socio-economic system is the second component that contributes to the perception of an impact on the local level. Depending on the position in the socio-economic system, the perception of the impact and the ability to cope with it might be quite different. Cutter's concept (Cutter 1996) of „hazard in place" was often seen as hazard in a specific geographic place, but Cutter stresses as well the social component including socio-economic structure, cognition of risk and ability to respond. These factors determine the social vulnerability of the place.

An example may illustrate the idea of a social space. In Pärnu, which was already mentioned above, parts of the city are not connected to the public water supply, but people get drinking water from private wells close to their houses. In Vana Pärnu (located west of the mouth of Pärnu river close to the shore) a part of the buildings is connected to the public water supply and a part is not. This results in a different impact depending on the kind of water supply and is rather independent of the geographic extent of SLR or a storm surge (Klein 2004). A number of houses is exposed to SLR in the same way, but the impact differs depending on the vulnerability. The location of the affected entity and the height of SLR in that place give information on the exposure, but for vulnerability (or maybe sensitivity talking about single objects) the characteristics of the entity are needed. However, aiming at adaptation strategies it might be useful to define (at least) two groups of affected. The one with public water supply faces generally lower impact and the responsibilities are mainly on the community level, i.e. Pärnu Vesi. For those with private water supply the impact is bigger and to a large part adaptation is in their own responsibility. So two different adaptation strategies with different stakeholders have to be developed, although it is the same geographic spot and in both cases the interaction between water supply and SLR/storm surge.

### c) Institutional context

The Pärnu case points already on a third criterion additional to the geographic and socio-economic context. Most impacts are related to the field of responsibility of certain institutions. Sometimes the responsibilities may be clearly divided, sometimes the competences of different institutions may overlap or sometimes there might be a lack of responsibility. Shakley & Deanwood (2002) suggest a „Frame of Reference". This concept is divided in a „Systems Frame of Reference", a „Process Frame of Reference" and an „Existing Responses Frame of Reference".

The „Systems Frame of Reference" is related mostly to a shared cognitive understanding and how experiences from past events can be utilized for climate change impact (see ASTRA winter storm case study, www.astra-project.org) As illustrated above, the experience of the winter storm in Pärnu led to a shared understanding and triggered some action at the highest level in town.

The „Process Frame of Reference" includes „existing ... institutions and processes, which are in place to manage resources and associated programmes, policies and projects." A positive example, where climate change fits into existing processes might be integrated coastal zone management (ICZM). In the ICZM for the Oder River (www.ikzm-oder.de) the consortium consists mainly of academic and research institutes, but the ICZM process includes cross sectoral and cross-national dialogs and a regional Agenda 21 group that ensures contact to local and regional actors. Climate change is explicit part of the ICZM for the Oder River (and should be according to the Recommendation of the European Parliament and of the European Council concerning the implementation of ICZM in Europe, chapter I (b) (2002/413/EC)).

In the Pärnu case, however it was difficult to find the right partners to discuss possible impacts of climate change, since the responsibilities are spread over many different institutions (harbour administration, water supply, town planning, nature conservation), but no procedure was on hand to coordinate a common strategy.

The last category „Existing Responses Frame of Reference" can be understood as a pool of known response options to the impact of climate change. Looking for tailor-made responses for climate change adaptation the list will be rather small, especially for the local level. However, there are other response options, which were not developed exclusively for climate change issues, but rather for other (often sectoral) problems. These can be understood as part of the institutional frame for climate change impact, since the responses are institutionalised, i.e. commonly and frequently used.

## 4. Conclusions

Based on the experience that setting climate change knowledge into a geographic context is not necessarily sufficient to get response on a local and regional level, it is suggested to set three context components: a *geographic space*, a *social* (or socio-economic) *space* and an *institutional space*. Obviously these spaces have strong interconnections and overlap, but they are not congruent. If all spaces are described sufficiently, the knowledge is well tied to the local or regional system. Also some questions can be related to each space:

- geographic space:
  - Were might be the impact?
  - How strong is the impact there?
- social space:
  - Who will be affected?
  - Which elements of the local/regional system will be affected?
- institutional space:
  - In which space of responsibility the impact takes place?
  - At which scale are the responsibilities (local, regional, national, EU, world)?
  - Are there any „windows of opportunity"?

Who should answer the questions? Most probably it has to be an iterative process. The ones presenting the scientific knowledge may have enough local insight to set the frame. Those affected at the local and regional level may not have sufficient awareness to estimate the whole extend of the impact. It has to be a deliberative process, where questions and answers are equally distributed among the participants. Ideally at the end of this process all relevant people are sitting at the same table and the development of adaptation strategies (adaptation as part of a policy evaluation according to Smit et al. 1999) can begin.

On first sight, this approach points very strictly to one direction. It tries to adjust information on climate change to a given system („Frame of Reference") and to keep necessary changes in the system minimal. On the other hand the iterative process to define the geographic, social and institutional space of climate change impacts itself raises awareness and set a basis for a network, as it was presented in connection to the SEAREG project by K. Schmidt-Thomé and Peltonen (2006). According to K. Schmidt-Thomé and Peltonen these four points are implicit for the term „network":

1. Social relations and collective action
2. Flexible organisation
3. Innovative linkages between entities that have not been understood as linked earlier
4. Shared agency in hybrid linkages between human and non-human actors (adopted from the actor-network theory)

Therefore, properly addressed and framed information can help to generate the necessary network that enhances the diffusion of knowledge and enables eventually the implementation of appropriate adaptation measures.

## References

*Bärring, L.* & *Persson, G.* (2006), „Influence of climate change on natural hazards in Europe", in Schmidt-Thomé, P. (ed.), Natural and Technological Hazards and Risks in European Regions, Special Paper 42, Geological Survey of Finland, Espoo, pp. 93–107.

*Cutter, S.L.* (1996), „Vulnerability to environmental hazards", Progress in human geography, vol. 20, no. 4, pp. 529–539.

*Eisenack, K.* & *Kropp, J.* (2006), „Regional Stakeholder's Perception of Climate Change in the Baltics", internal project document.

*Fleischhauer, M.* (2004), „Klimawandel, Naturgefahren und Raumplanung – Ziel- und Indikatorenkonzept zur Operationalisierung räumlicher Risiken", Dortmunder Vertrieb für Bau- und Planungsliteratur, Dortmund.

*Greiving, S.* (2004), „Risk Assessment and Management as an Important Tool for the EU Strategic Environmental Assessment", DISP, vol. 157, pp. 11–17.

Ilmatieteenlaitos (2006), "Kesän 2006 säät, kuivuus ja lämpö", available www.fmi.fi/saa/tilastot_161.html, visited 05.02.2007.

IPCC (2007), „Climate Change 2007: The Physical Science Basis – Summary for Policymakers", Contribution of Working Group I to the Fourth Assessment Report of the Intergovernmental Panel on Climate Change, available http://ipcc-wg1.ucar.edu/wg1/docs/WG1AR4_SPM_Approved_05Feb.pdf.

*Klein, J.* & *Staudt, M.* (2006), „Evaluation of future sea level rise impacts in Pärnu (Estonia)". In Schmidt-Thomé, P. (ed.), Sea Level Change Affecting the Spatial Development in the Baltic Sea Region, Special Paper 41, Geological Survey of Finland, Espoo, pp. 71–82.

*Klein, J.* (2004), „GIS-based impact assessment of Sea Level Rise on the City of Pärnu (Estonia) – Application of three Sea Level Rise Scenarios on a Local Level", final thesis, University of Stuttgart.

*Kussisto, E.* (ed.) (2006), „Finland's Fourth National Communication under the United Nations Framework Convention on Climate Change", Tampere.

*Lehtonen, S.* & *Peltonen, L.* (2006), „Risk communication and sea level rise: bridging the gap between climate science and planning practice". In Schmidt-Thomé, P. (ed.), Sea Level Change Affecting the Spatial Development in the Baltic Sea Region, Special Paper 41, Geological Survey of Finland, Espoo, pp. 61–70.

*Schmidt-Thomé, P.* (2006), „Integration of natural hazards, risk and climate change into spatial planning practices", Academic dissertation no. 193, University of Helsinki, Geological Survey of Finland.

*Schmidt-Thomé, K. & Peltonen, L.* (2006), „Actors, networks and actor-networks in coping woth sea level rise". In Schmidt-Thomé, P. (ed.), Sea Level Change Affecting the Spatial Development in the Baltic Sea Region, Special Paper 41, Geological Survey of Finland, Espoo, pp. 51–59.

– (2006), „Sea level Change Assessment in the Baltic Sea Region and Spatial Planning Responses". In Schmidt-Thomé, P. (ed.), Sea Level Change Affecting the Spatial Development in the Baltic Sea Region, Special Paper 41, Geological Survey of Finland, Espoo, pp. 7–17.

*Shackley, S. & Deanwood, R.* (2001), „Stakeholder perceptions of climate change impacts at the regional scale: implications for the effectiveness of regional and local response", Journal of Environmental Planning and Management, vol. 45, no. 3, pp. 381–402.

*Smit, B., Burton, I., Klein, R.J.T. & Street, R.* (1999), „The science of adaptation – a framework for assessment", Mitigation and Adaptation Strategies for Global Change, vol. 4, pp. 199–213.

*Virkki, H., Kallio, H. & Orenius, O.* (2006), „Sea Level Rise and Flood Risk Assessment in Itä-Uusimaa". In Schmidt-Thomé, P. (ed.), Sea Level Change Affecting the Spatial Development in the Baltic Sea Region, Special Paper 41, Geological Survey of Finland, Espoo, pp. 95–106.

# Chapter 3
# BALTEX Assessment of Climate Change for the Baltic Sea Basin (BACC)

**Hans von Storch[1/2] / Anders Omstedt[3] / Hans-Jörg Isemer[4]**


## Abstract

A voluntary group of about 80 scientists from 12 countries has reviewed and assessed the published literature on climate change in the Baltic Sea basin; here, the term „climate change" refers to all changes of weather statistics, in the atmosphere, Baltic Sea and cryosphere, whether they are related to human action or due to natural variability. After two years of work, which includes external and internal review, a conclusion has been arrived at – in almost all aspects, this conclusion represents a common assessment; only in minor aspects dissent was found and documented. The overall assessment will be published in a book later in 2007. The conclusions of the assessment have been accepted by HELCOM, the Baltic Marine Environment Protection Commission, as a suitable analysis of the state of knowledge. In this short chapter, the key findings are summarized and illustrated with a few key diagrams on past and expected future developments. Also, the background of the BACC initiative is described.


## Background of BACC

The *purpose* of the BACC assessment is to provide the scientific community and the public with an assessment of ongoing and future climate change in the Baltic Sea basin – that is the Baltic Sea water body and the catchment area. This is done by reviewing the scientific literature and the published knowledge about climate change in the Baltic Sea basin. The BACC Project integrates available knowledge of historical, current and of expected future climate change. Similar to the IPCC process, BACC is not pursuing its own scientific analysis but is merely reviewing what is „out there".

An important element of the BACC assessment is the comparison with the historical past (until about 1800) to provide a framework for the severity and unusualness of the change. Another important aspect of BACC is the combination of evidence on climate change and related impacts on marine, freshwater and terrestrial ecosystems in the Baltic Sea basin.

---


1 Institute for Coastal Research, GKSS Research Center, Germany, Max-Planck Straße 1, 21502 Geesthacht, hvonstorch@web.de.

2 Zentrum für Marine und Atmosphärische Wissenschaften, Hamburg, Germany.

3 Göteborg University, Sweden.

4 International BALTEX Secretariat, GKSS Research Center, Germany.

BACC addresses the problem of „climate change", which is unfortunately differently understood in different quarters (cf. Bärring 1993). The problem is that „inconstancy" is an inherent property of the climate system. The United Nations Framework Convention on Climate Change (UN-FCCC) defines climate change as „a change of climate that is attributed directly or indirectly to human activity that alters the composition of the global atmosphere, and that is in addition to natural climate variability over comparable time periods." By contrast, the Intergovernmental Panel on Climate Change (IPCC) defines climate change broadly as „any change in climate over time whether due to natural variability or as a result of human activity." Obviously, it is rather important which definition is used, in particular when communicating with the public and the media. BACC has decided to essentially follow the IPCC definition, i.e., „any change in climate over time whether due to natural variability or as a result of human activity."

BACC is the first systematic scientific effort for assessing climate change in the Baltic Sea basin. More than 80 scientists from 12 countries have contributed on a voluntary basis. The results have not been influenced by either political or special interests.

BACC was „invented" during a BALTEX (Baltic Sea Experiment)[5] meeting in Roskilde in September 2004, installed on an ad-hoc basis by selecting a steering group in November 2004, and later formally accepted and endorsed by BALTEX as a BALTEX project[6]. At the same time, it was agreed that HELCOM would consider the BACC assessment as a basis for its own assessment. This cooperation between HELCOM and BALTEX as a major research programme of the World Climate Research Programme is unique and of mutual benefit.

After having formed a group of lead authors, who are responsible for one of the four chapters and after having identified the contributing authors in a bottom-up approach, draft assessments were established. The BACC draft material was reviewed internally at two meeting in Helsinki and Warszawa in 2005. An external review followed, organized under the guidance of the BALTEX SSG. The publication of the „BACC Book" (BACC Group 2007) with the summary for policymakers, the detailed assessment chapters and a number of annexes explaining technical background and terminology is presently being completed.

In this chapter, we first present the official summary of the key findings of the assessment, agreed among the chairs of BACC[7] and its lead authors[8].

---

5 See: www.baltex-research.eu

6 See: www.baltex-research.eu/BACC

7 Hans von Storch & Anders Omstedt.

8 Raino Heino, Heikki Tuomenvirta, Bo Gustafsson and Valery Vuglinsky for the chapter „Past and Current Climate Change, Detection and Attribution"; Phil Graham for chapter „Projections of Future Climate Change"; Benjamin Smith for the chapter „Climate-

## Summary statement

In the past century here has been a marked increase of *temperature* of more than 0.7°C in the Baltic region, which is larger than the global mean temperature increase. Consistent with this increase in mean and extreme temperatures, other variables show changes, such as increase of winter runoff, shorter ice seasons and reduced ice thickness on rivers and lakes in many areas. These trends are statistically significant but they have not been shown to be larger than what may be expected from natural variability. In addition, no robust link to anthropogenic warming, which on the hemispheric scale has been causally related to increased levels of greenhouse gases in the atmosphere in „detection and attribution" studies, has been established. However, the identified trends in temperature and related variables are consistent with regional climate change scenarios prepared with climate models. Therefore, it is plausible that at least part of the recent warming in the Baltic Sea basin is related to the steadily increasing atmospheric concentrations of greenhouse gases. Efforts are needed which systematically examine the inconsistency of recent trends with natural variability, circulation changes as well as the consistency with elevated greenhouse gas concentrations as a potential cause.

The situation is much less clear regarding *precipitation*: In the past, a spatially non-uniform pattern of upward and downward trends has been observed, which can hardly be related to anthropogenic climate change. For the future, intensified winter precipitation may emerge later in this century over the entire area, while summers may become drier in the southern part – but this expectation is uncertain for the time being. For the water body of the Baltic Sea, a tendency towards lower *salinity* is expected. Similarly, no clear signals, whether for the past or for the scenarios, are available with regard to *wind* conditions.

In view of the large uncertainty in our knowledge about the changing climatic conditions, it is not surprising that knowledge about ecological implications of ongoing and future climate change is far from complete and also very uncertain. The observed changes in temperature in the past have been associated with consistent changes in *terrestrial ecosystems*, such as earlier spring phenological phases, northward species shifts and increased growth and vigour of vegetation. In lakes, higher summer algal biomasses have been found. These trends are expected to continue into the future; induced species shifts may be slower than the warming which causes it. In the *marine ecosystem* of the Baltic Sea, the assessment is particularly difficult because of the presence of strong non-climatic stressors (eutrophication, fishing, release of pollutants) related to human activities. Changing temperatures have been related to various effects, in particular to the composition of species. A lowering of salinity is thought to have a major influence on the distribution, growth and reproduction of the Baltic Sea fauna.

---

related Change in Terrestrial and Freshwater Ecosystems"; and Joachim Dippner and Ilppo Vuorinen for chapter „Climate-related Change in Marine Ecosystems".

Freshwater species are expected to enlarge their significance, and invaders from warmer seas (such as the zebra mussel *Dreissena polymorpha* or the North American jelly comb *Mnemiopsis leidyi*) are expected to enlarge their distribution area. The expected changes in precipitation (and thus river runoff) may have additional detrimental effects on the problem of eutrophication.

## Illustrations

In the following some of the key findings of past developments and possible future developments (scenarios) are illustrated with a few diagrams; further details are provided in the forthcoming BACC book (BACC Group 2007).

1. Figure 1 shows the development of annual mean surface air temperatures in the Baltic Sea basin, namely a steady increase in all areas. The temperatures at the end of the 20th century are more than 1°C warmer than at the beginning of the record in 1871.

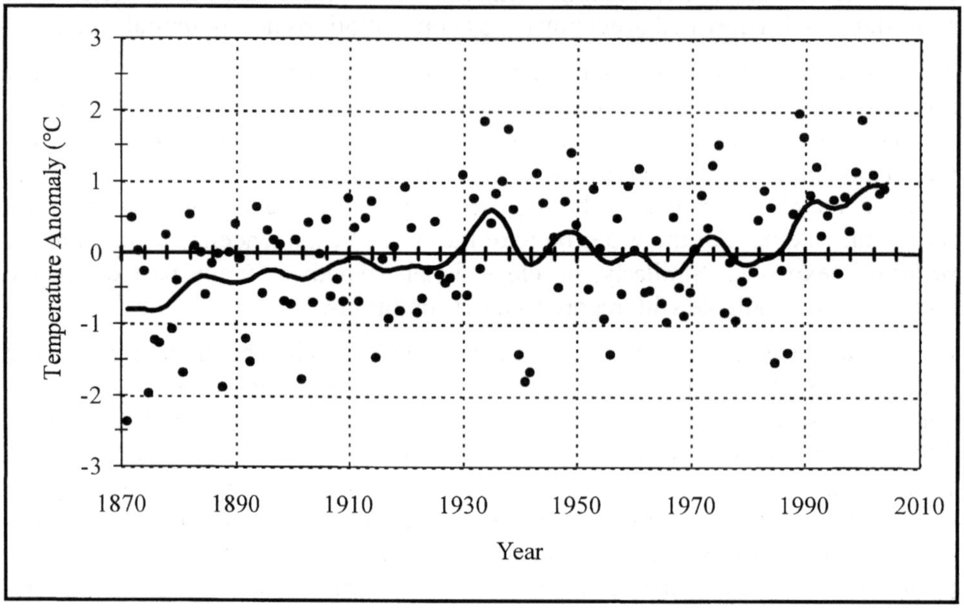

2. The warming is also reflected in increased run-off from rivers in winter – as is exemplified by the run-off from the Finnish territory into the Baltic Sea in February, displayed in Figure 2.

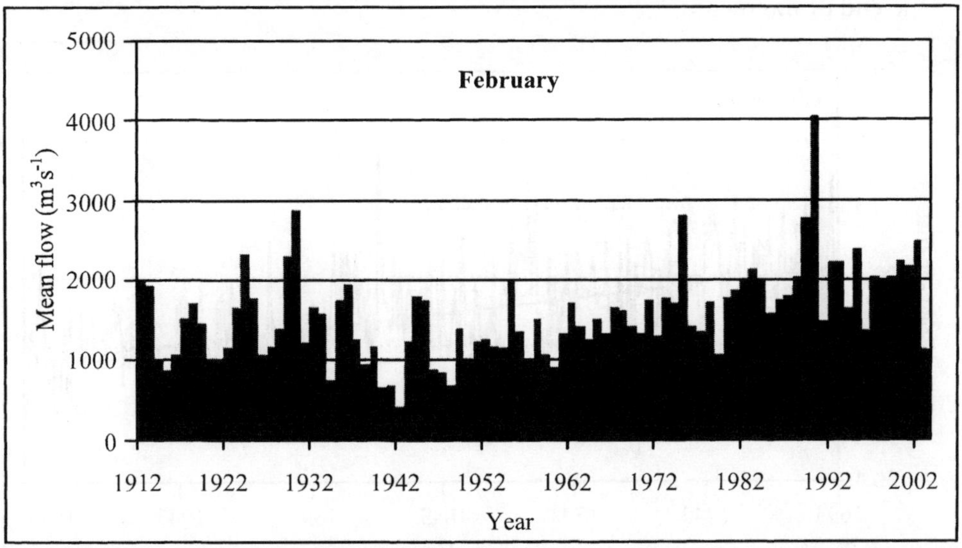


3. Figure 3 shows the duration of ice coverage of a Russian lake since 1946 – the duration is shortened by more than 25 days since then.

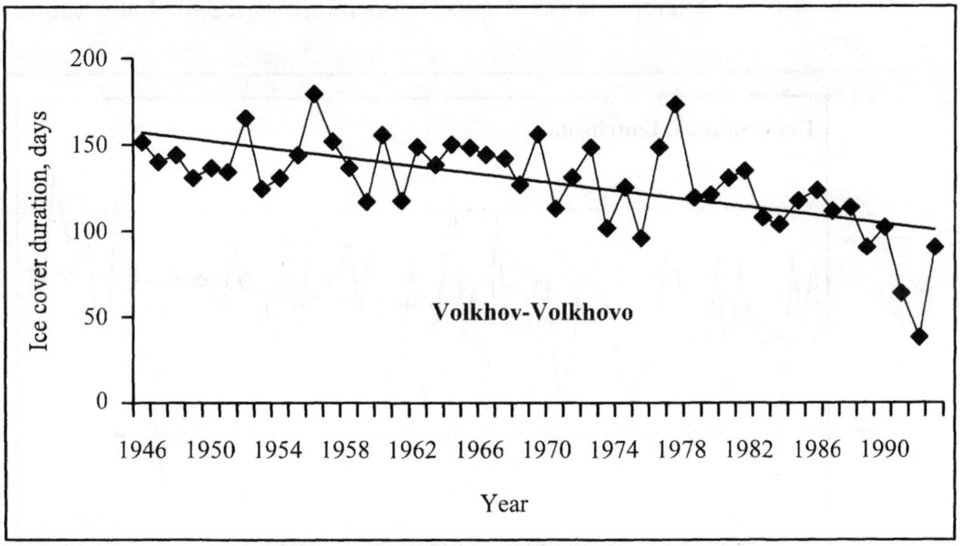

4. That this may be part of a longer development is demonstrated by Figure 4, which shows the date of ice break up in a river in northern Finland, which is documented since 1693. Interestingly, the trend does not seem to accelerate at the end of the record.

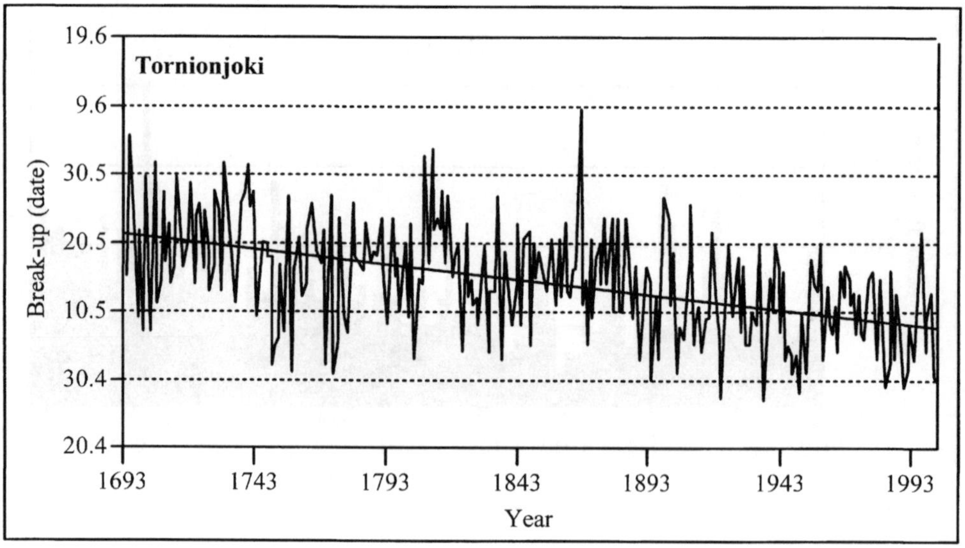


5. At the same time, summer water temperatures in lakes, for instance in Finland, rose. An example is the August temperature series for Lake Saimaa in Figure 5.

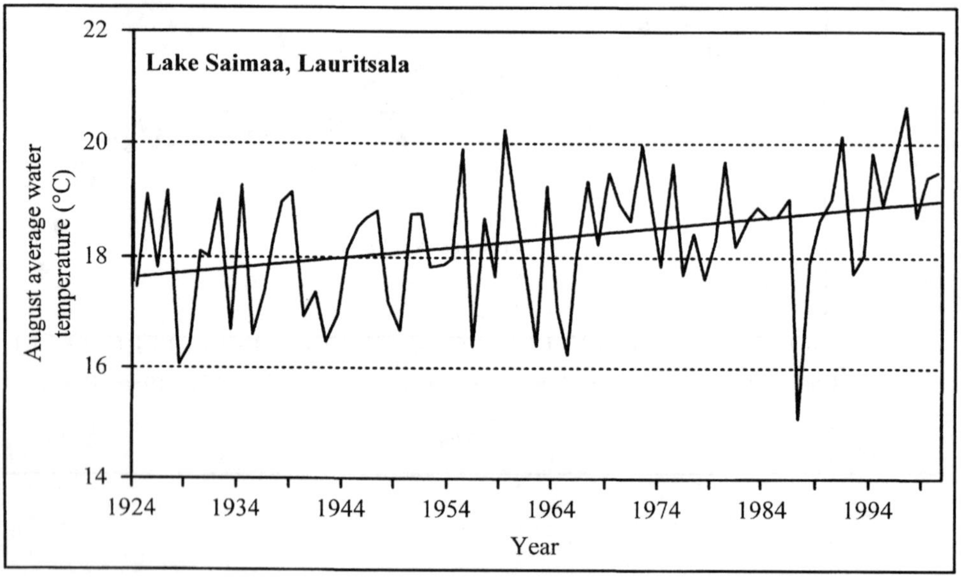

6. Seasonal precipitation ratios between the two 25-year periods 1976–2000 and 1951–1975 based on VASClimO data (Figure 6; Beck et al. 2005). The southern part became somewhat drier in summer (JJA) in the second 25 years, while the north was in many parts drier in winter (DJF) than earlier.

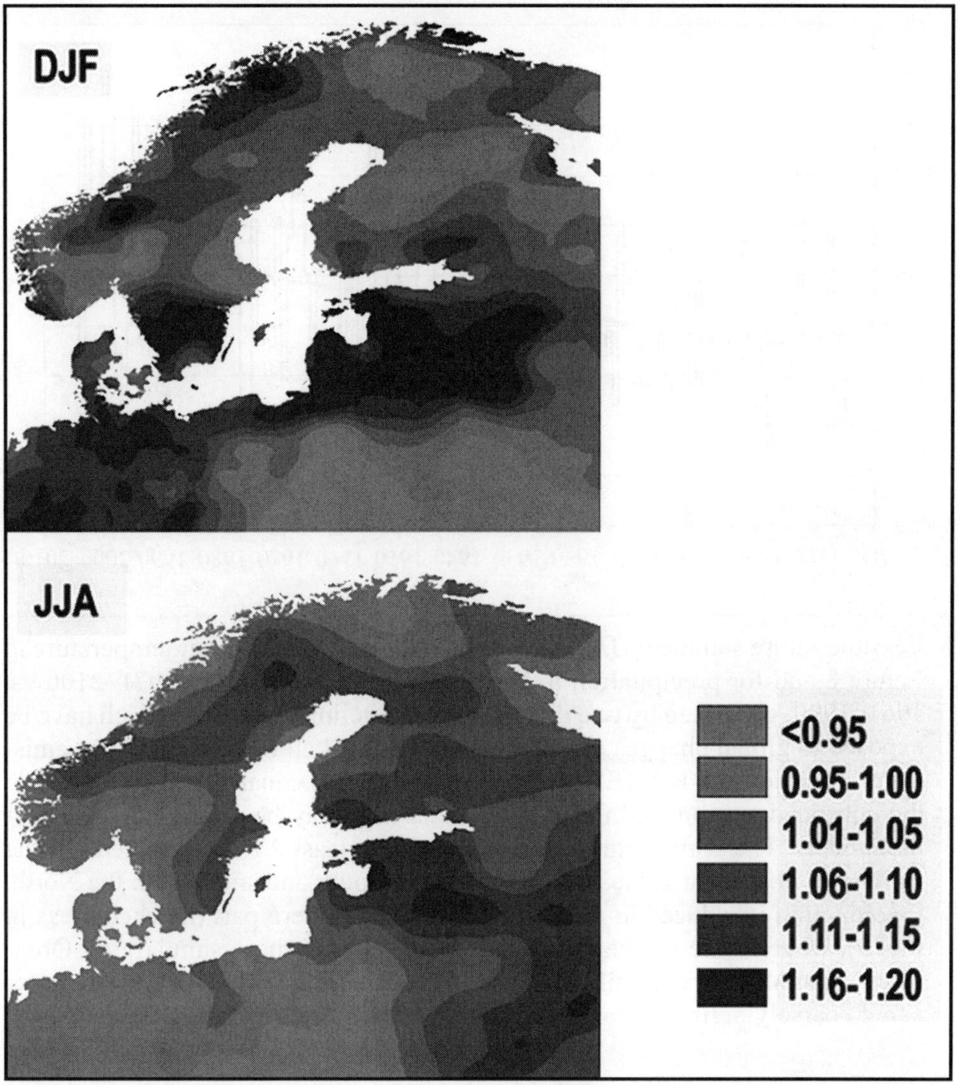

7. The annual precipitation amount over Denmark saw, after suitable smoothing, a steady increase by about 1/6 during the instrumental period 1874–2004 (Cappelen & Christensen 2005; Figure 7).

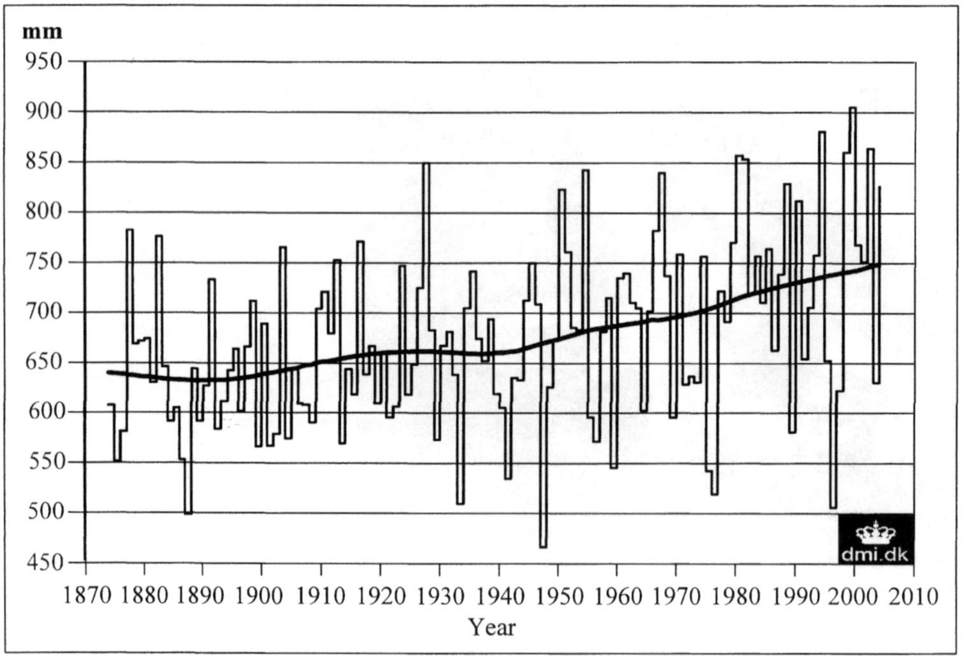


8. Possible future summer (JJA) states are displayed for surface air temperature in Figure 8 and for precipitation in Figure 9. Shown are changes 2071–2100 vs. 1961–1990, generated by two different regional climate models, which have be exposed to global changes generated by two global climate models. The emissions scenario used is SRES A2. All four maps are equally probable, even if the enhanced warming of the Baltic Sea surface may be an artifact of one of the models. The entire region warms up with at least 2°C until the end of the century, with largest values (up to 7°C) in the south and smallest in the North. Precipitation is reduced in all four cases in the southern part (hatched areas in Fig. 9), whereas the northern part becomes wetter in these simulations. Obviously, spatially very detailed statements make little sense; only statements about coarsely defined areas are consistent across different scenarios.

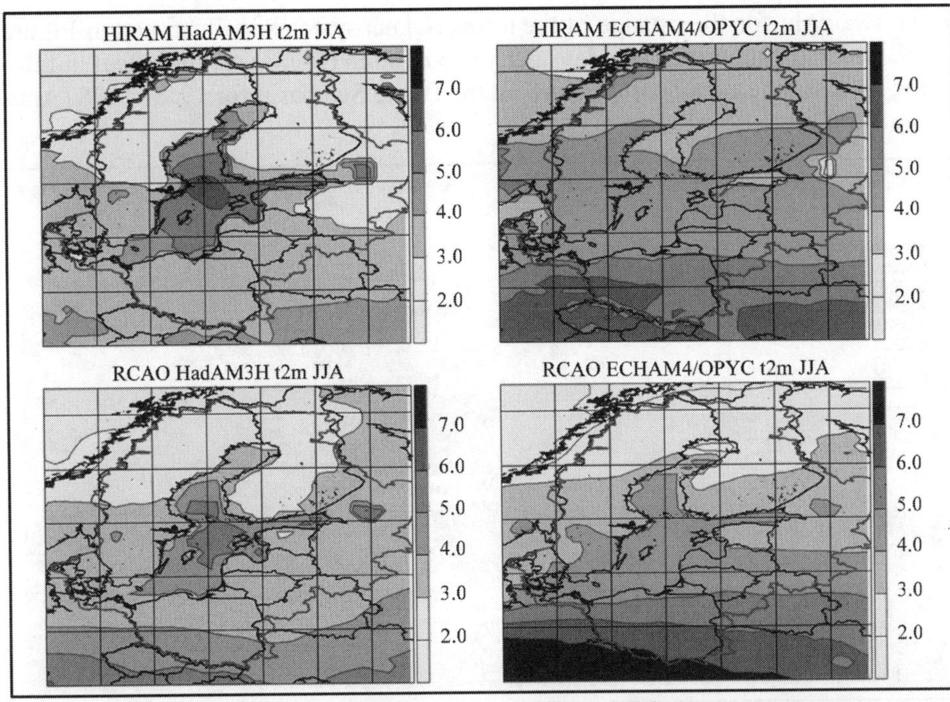
HIRAM HadAM3H t2m JJA
HIRAM ECHAM4/OPYC t2m JJA
RCAO HadAM3H t2m JJA
RCAO ECHAM4/OPYC t2m JJA
7.0
6.0
5.0
4.0
3.0
2.0

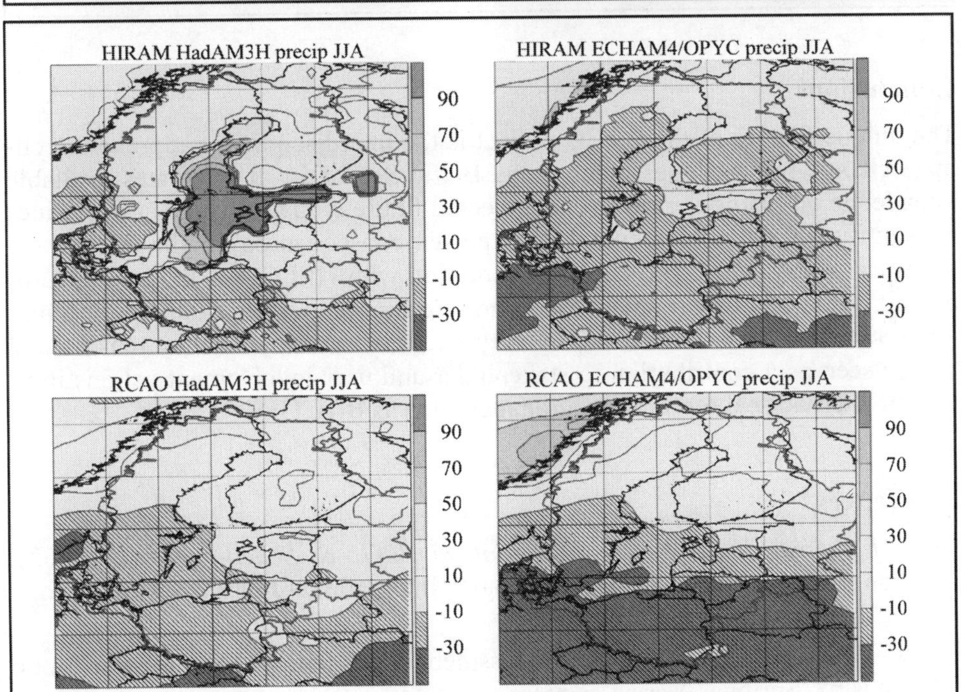
HIRAM HadAM3H precip JJA
HIRAM ECHAM4/OPYC precip JJA
RCAO HadAM3H precip JJA
RCAO ECHAM4/OPYC precip JJA
90
70
50
30
10
-10
-30

9. An example for the change of the terrestrial ecosystems is illustrated in Figure 10, which displays the rate (days per year) in birch (*Betula pendula*) leaf unfolding over southern and eastern parts of the Baltic Sea basin for 1951–1998 (Ahas et al. 2002).

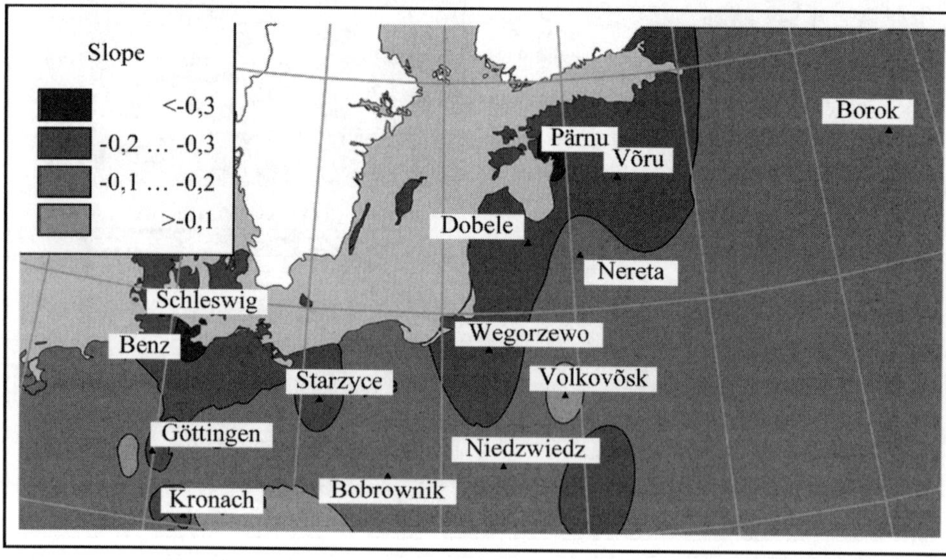


## Conclusions

The unique feature of the BACC project is the interdisciplinary approach to climate change impact assessment in the Baltic Sea basin. It integrates available knowledge on historic, current and expected future climate change and provides a scientifically sound perception on expected consequences for the marine and terrestrial ecosystems. The close collaboration with HELCOM, the intergovernmental Baltic Sea Environment Protection Commission, is an example for a dialogue between the scientific community and environmental policy makers and can be seen as a contribution to stakeholder and user involvement actions initiated in international research programmes such as BALTEX.

## References

*Ahas, R., Aasa, A., Menzel, A., Fedotova, V.G. & Scheifinger, H.* (2002), „Changes in European spring phenology", International Journal of Climatology 22: 1727–1738.

BACC Group (2007), „BALTEX assessment of climate change for the Baltic Sea basin", Springer-Verlag, in press.

*Bärring, Lars* (1993), „Climate – change or variation?" Climatic Change 25: 1–13.

*Beck, C., Grieser, J. & Rudolf, B.* (2005), „A New Monthly Precipitation Climatology for the Global Land Areas for the Period 1951 to 2000", published in Climate Status Report 2004, pp. 181–190, German Weather Service, Offenbach, Germany.

*Cappelen, J. & Christensen, J.H.* (2005), „DMI's bidrag til Danmarks 4. nationale afrapportering til FN's klimakonvention UNFCC". Danmarks Klimacenter rappport 05–05.

Chapter 4

# Promoting Climate Friendly Communities in Northamptonshire, United Kingdom

**Philip Gray[1], Shane Conway[2]**


## Abstract

The Climate Friendly Communities Northamptonshire (CFC-N) project aims to encourage and support communities to reduce their environmental impact, and notably their carbon dioxide emissions, which is under development by a partnership of local authorities and organisations in Northamptonshire, England. The aim is to motivate efforts by geographical communities (e.g. villages or areas within towns) to assess their current carbon emissions, identify ways to reduce them and agree targets and actions. The national background includes raised levels of concern about climate change in the UK, and an increased focus on the role of local government in achieving this. The key supportive processes are government incentives for public agencies to work together (Local Area Agreements), and the voluntary Nottingham Declaration on Climate Change for local authorities. Another factor is the existence of local government energy efficiency programmes. Various other community-level schemes exist around the UK, which provide inspiration for aspects of this project. Unlike most other projects, this one seeks to measure baseline emissions and actual progress, and to encompass emissions from all sources including domestic, business and other organisations (e.g. schools). The scheme aims to engage about seven communities in the first year and to encourage future progress through a graded „award" scheme. The project demonstrates the value of a „climate change declaration" and a multi-agency approach. By mid-2008, the project should provide information about the usefulness of a community-based approach, and initial data about the prospects for reducing actual carbon emissions in real-life communities in the UK.


## Communication about climate change in the United Kingdom

Much communication about climate change, including mass communication, is targeted at individuals. While this kind of communication has been very successful in recent years at raising awareness of the issues, there is evidence to indicate that it is not very successful in changing behaviour. It has been reported, for example, that a recent study of public attitudes to energy and the environ-


1 Northamptonshire County Council, England, UK.

2 Northampton Borough Council, England, UK.

ment showed that while readers of the liberal newspapers *The Guardian* and *The Independent* were among the most concerned about climate change, they were no more likely to have taken any specific energy saving measure, and actually less likely to have insulated their homes (EIBI 2007, reporting a University of Cambridge survey).

More targeted communication campaigns based on personalised marketing are known to be much more successful in changing environmental behaviour (Sustrans, 2006), but so far have mainly been used in the context of transport choice.

A key field for practical action on climate change is domestic energy efficiency. Communication about energy efficiency in the United Kingdom (UK) is based mainly on a mass „stimulation of demand" model. Marketing seeks to raise individual awareness about the reasons to save energy, and about where to obtain advice (typically, the local or regional energy advice service, or a national body such as the Energy Savings Trust). These services then respond to individual requests arising from this marketing. In addition more targeted campaigns are carried out with particular sectors of society, notably the elderly and people on low incomes, who are identified as a priority for improving domestic conditions. Communication on this latter aspect is promoted nationally by the charity NEA (National Energy Action).

## Community approaches to climate change

This paper examines a different approach to energy efficiency and climate change advice, based on community action. The central idea is that communication campaigns aimed at changing individual behaviour are not sufficient, and that we also need to harness social processes in the form of community-based action, if real and widespread change is to be achieved.

The concept of local communities acting to tackle global environmental challenges is, of course, not new, and was the basis of the Local Agenda 21 (LA21) efforts that followed the Rio Earth Summit in 1992. Local government in the United Kingdom was an early and enthusiastic supporter of LA21, which appeared to offer a way of taking environmental action in spite of significant financial and legal constraints on local government at the time. The wave of activity on LA21 reached its peak in the late 1990s and has now largely run its course.

The mid-2000s has seen renewed interest among British people in what individuals can do to reduce their environmental impact (DEFRA 2006) in the face of sustained scientific activity and media interest in climate change (e.g. Met Office 2006, 2007).

At the same time, central government has explicitly acknowledged that local government has the potential to play a critical role in achieving the UK's climate change goals (DCLG 2006). In addition, there is a strong awareness in government and its agencies that domestic carbon emissions make a significant contribution to the UK's greenhouse gas inventory, and are one of the areas with the biggest potential reductions. Another important factor is that central government is also encouraging local government authorities, other public bodies and central government agencies to work together more effectively to tackle issues on the local level.

Another important driver is the Nottingham Declaration, which is a voluntary pledge for local authorities in the UK to take action on climate change[3]. This declaration was developed in Nottingham, UK in 2000 and spread „horizontally" among local authorities, before being re-launched with a higher profile in 2005. Since then the UK government has also recognised its value and encouraged local authorities to sign up. All the authorities in Northamptonshire signed up by October 2005, thanks to a specific to increase sign-up in the East Midlands Region. This has encouraged the authorities to look at ways to implement the Declaration jointly, such as the Climate Friendly Communities project.

The authorities involved in this project are responding to these drivers by working jointly to encourage geographical communities in Northamptonshire to come forward and commit themselves to reducing their climate impact.

The project builds on the 'Carbon Neutral' approach pioneered by the village of Ashton Hayes in Cheshire (a community/parish council led initiative), as well as the countywide initiative supported by the University of East Anglia in Norfolk, known as CRed Carbon Reduction Project. It also draws on the work of other regional partnerships such as the South East Climate Change Partnership.

It is hoped that using a community-based approach, it will be possible to involve households that would not be reached by a „demand stimulation" approach. Instead of relying on people responding to „vertical" marketing, communities will be able to utilise more natural, „horizontal" and local links. Ideally there will be positive social reinforcement for taking part, instead of people having to be innovative or different from their neighbours. In addition there ought to be a positive effect by concentrating action in one area at a time, so that a higher proportion of people are exposed to the project and its outcomes. Finally, success stories about communities participating in the project can be used to provide an example for others in the future.

---

3 www.nottinghamdeclaration.org.uk.

## Outline of the Climate Friendly Communities Northamptonshire Initiative

Climate Friendly Communities Northamptonshire (CFC-N) is a project to encourage and support communities in Northamptonshire to reduce their environmental impact, with a particular focus on carbon dioxide emissions, which is being developed by a partnership of local authorities and other organisations in Northamptonshire, England. The aim is to motivate geographical communities (e.g. villages or areas within towns) to take action to understand and reduce their total carbon emissions in a measurable way.

The emphasis on measurement sets this project apart from some of the other approaches mentioned above. The aim is to produce a baseline „carbon profile" for each community in the programme, analogous to the national and even district-level carbon emission profiles now produced in the UK and other countries. These profiles will be based on the energy bills and travel details of households along with the energy and other consumption details from the local school(s) and businesses.

Once the baseline has been established, actions and targets for reducing emissions will be agreed between the community and the project advisers.

## Partners

The project is led by the Northamptonshire Energy Efficiency Partnership (NEEP), which represents the officers responsible for implementing energy efficiency legislation in each district in the county. A working group is developing the project concept, consisting of representatives from NEEP and co-opted members from the county council, the county's climate change partnership, the University of Northampton and a rural development organisation, Northamptonshire ACRE (Action for Communities in Rural England).

The project will be launched centrally by NEEP with a publicity campaign and a launch „open day" where advice and support will be on offer. NEEP will also select the communities to take part, based on their applications. The responsibility for baseline data collection and processing will be shared between the university partners and other partners currently being identified. The local energy efficiency team in each district will support the communities in setting their targets, agreeing an action plan and implementing it – for example with advice about practical actions to reduce home emissions, as well as information about grants and other support.

NEEP will monitor progress and coordinate reviews of progress by each community, and provide overall support to the project.

## Project aims and criteria

The project has two key aims. The first is to raise awareness among the members of community of the sources and amounts of carbon emissions from their community, and how they can reduce them.

The second aim is to encourage all residents/local businesses to adopt measures to reduce carbon emissions, in particular by increasing the uptake of the (existing) schemes promoting energy efficiency improvements across the county, and by adopting renewable and low carbon energy technologies.

A Climate-Friendly Community is envisaged as one that is working towards, or actually, reducing its total carbon dioxide emissions (and taking other actions to reduce its environmental impact). Actions to adapt to climate change will also be encouraged as the project develops, but will not be stressed in the initial marketing in order to avoid any possible confusion about the project's aims[4].

The „communities" could be urban or rural and might consist, for example, of a village, a street or an area within a town. Whatever the type of area, there needs to be an organisation in place, such as a residents' association, parish council, a regeneration board or a community group, which is willing to lead the process in the community.

A community, which registers and shows some progress over the first year towards agreed objectives to reduce emissions, will be designated a „Climate Friendly Community" at a certain level (e.g. „Bronze" level). Further progress in later years will lead to a higher level designation (e.g. „Silver" or „Gold"). The names and exact criteria for different levels are still under development, but will draw upon existing schemes such as the Eco-Schools and Healthy Schools programmes, which employ similar rating systems.

The initial aim will be to engage one community in each of Northamptonshire's seven districts in the programme during the first year. Each of these seven communities should have produced its carbon profile and agreed an action plan by the end of that first year.

## Marketing and selection of communities

The scheme will initially be marketed during Spring 2007 through general advertising using leaflets, local media and websites, along with targeted advertis-

4 The project partners believe that it would be premature to emphasise adaptation at present, since there is still only very limited understanding of that concept in the public realm. Discussing adaptation at this point could lead to unproductive debates about the relative merits of mitigation and adaptation. However the project may well offer opportunities to develop an understanding of adaptation in the participating communities.

ing to organisations or areas, which are believed to have a good chance of being interested and effective candidates. These include areas with residents' associations, parish councils, various community organisations, areas involved in regeneration programmes and areas with an Eco-School[5]. The places targeted will include a balance of urban and rural-based areas.

The advertising will invite organisations to apply for an information pack, which will give guidance on submitting an expression of interest (with a provisional deadline of the end of April 2007).

Expressions of interest in this first round will be assessed according to the prospects of successful participation, and possibly the degree of impact to be expected. Some allowance may be made for the social benefits expected, e.g. by improving conditions for elderly residents. Specific factors will include evidence of leadership and enthusiasm to take part, any past success in community change, and creative ideas about possible activities to reduce emissions (regardless of how feasible or well informed these appear). The level of prior knowledge about the issue will not be a key factor.

The communities selected will be sent a final information pack with further guidance on taking part.

## Launching the project

The active phase of the programme will be publicly launched at a World Environment Day exhibition to be held in Northampton, the county town of Northamptonshire. This will include workshops for participants as well as the opportunity to meet agencies, groups and companies offering specific advice and services, for example on environmental technologies.

Each participating community will be expected to establish a partnership to develop their own 'Climate Friendly Community Plan'. This partnership should if possible involve a range of different sectors, including the town or parish council (if one exists), the local school(s) and local businesses.

## Setting the baseline

The first practical step for each community will be to produce their own carbon profile. These should provide a concise summary of the emissions due to different activities in the community (including domestic energy use, business energy use, waste disposal and recycling, local transport, national/international trans-

---

5 Eco-Schools is a national programme to encourage and enable schools to become more sustainable. See www.eco-schools.org.uk/index.htm.

port and use of materials). Information about the average emissions for the district or county and for the UK overall will also be provided for comparison.

The detailed procedure for producing the profiles is still under development at the time of writing, but it will build upon existing software such as the South East Climate Change Partnership's Emissions Calculator (SECCP 2006). The project team intends to make a software tool available to the communities for self-completion, but it is expected that most will require some help from NEEP to gather and process the data for their whole community.

An appropriate sampling method will also be developed with advice from partners, in order to guide the communities, and enable robust comparisons over time.

In the absence, as yet, of any nationally agreed methodology, especially in relation to the potential for 'offsetting' carbon emissions by planting trees and the absorption of $CO_2$ from the atmosphere by existing trees, the assumptions made in estimating carbon emissions will need to be reviewed and revised from time to time.

Although the main focus will be on carbon dioxide emissions and climate change, communities will also be encouraged to work on other environmental issues (e.g. waste reduction, recycling and treatment) where there are particular interests or opportunities to do so. Adaptation activities – e.g. urban tree planting to reduce heat island effects – will also be taken into account.

## Support and incentives

Support will be available from the county partnership and local energy efficiency officers. This will include a training programme so that each community can collate the information needed for their own community carbon profile.

The local authorities will act as mentors, but the drive and leadership for activities will need to come from members of the communities involved. The local authorities are however also seeking to lead by example, ensuring that their direct activities become more climate friendly, both in any activities they carry out within the participating communities, and generally across the district or county.

One of the ways we will help communities achieve their goals will be to provide information and education about issues relating to recycling, energy and any other aspects of sustainable living.

Two forms of incentive will be provided, over and above the inherent motivation of helping to combat climate change. First, a community will be rewarded for progress over time by being designated as a Climate Friendly Community at a specific level. There will be the opportunity to progress to a higher level in future years. Second, it is hoped to recognise the communities that make the most progress in a given year with an award, if possible supported by prizes, such as solar panels for the local school. This is intended to promote competition between communities and maximise the publicity impact of the programme.

## Assessing progress

Participating communities will need to complete an application form for „Climate Friendly" status each year, which will include details of their achievements over the last year and their plans for the next year. This will be backed up by an updated carbon profile, which will be subject to quality control e.g. through a selective auditing process, or possibly being carried out directly by NEEP.

The applications will be judged by a panel of at least two independent experts, who will award „Climate Friendly" status at a given level for the year, as well as judging eligibility for any prizes, based on a combination of the progress achieved and the quality of the future plans.

## Conclusions

This project is a work in progress, and there are no final results to report at this stage. However some preliminary conclusions can be drawn about the factors that are conducive to such projects coming into being.

1. *Development of a national consensus on the reality and importance of climate change.* Such a consensus has developed to a dramatic degree over the last two or three years in the UK. There is no doubt that without such a consensus it would be more difficult for public authorities to launch projects of this kind.
2. *Promotion of partnership working among public agencies by central government.* The gradual development of a culture of cooperative working, thanks to direct incentives by central government, has greatly helped this project to overcome institutional barriers more easily.
3. *Recognition of the local dimension of climate change.* Both local government (e.g. in the form of the Nottingham Declaration on Climate Change) and more recently, central government, have recognised that many of the key carbon-producing activities are influenced, often significantly, by local authorities. This has greatly helped to create the space for local authorities to work with their communities on this issue.
4. *Existence of an energy efficiency promotion infrastructure.* This infrastructure of public energy efficiency advice services is seen as providing the practical „backbone" for this project. National bodies like the Energy Saving Trust and National Energy Action also have an important background support role.
5. *Recognition of the community dimension of communication on climate change.* Pioneering communities like Ashton Hayes in Cheshire have helped to spread awareness that it is not sufficient to rely on individual decisions to tackle the

carbon challenge, and that the community dimension is a (more) promising direction for future communication on climate change. Local Agenda 21 can also be seen as having contributed to this development.

6. *Existence of specific resources such as carbon profiling tools*. Although there is not a single, readymade tool that completely meets the needs of this project, there are plenty of examples that can be drawn upon, which avoids the need to start from scratch.

## References

DCLG (Department for Communities and Local Government) (2006), Strong and Prosperous Communities – Local Government White Paper, vol. II – Annex F: Climate Change. Cm 6939-II. HMSO, Norwich, October 2006.

DEFRA (Department for the Environment, Food and Rural Affairs) (2006), Attitudes to Climate Change – Wave 4. Topline Summary. Prepared for DEFRA by COI (Central Office of Information), London, September 2006.

EIBI (Energy in Buildings and Industry) (2007), People thinking about but not acting on environment – survey reveals divergence between action and attitude. Energy in Buildings and Industry, January 2007, p. 10.

Met Office (2006), „Exceptionally warm extended summer 2006". News release, 16 October 2006. UK Meteorological Office, Exeter, United Kingdom. Available www.metoffice.gov.uk/corporate/pressoffice/2006/pr20061016.html

– (2007), „2007 – forecast to be the warmest year yet". News release 4 January 2007. UK Meteorological Office, Exeter, United Kingdom. Available www.metoffice.gov.uk/corporate/pressoffice/2007/pr20070104.html

SECCP (South-East Climate Change Partnership (2006), SECCP Emissions Monitoring Toolkit. SECCP, Guildford, Surrey. Available www.climatesoutheast.org.uk ⇨ Publications.

Sustrans (2006), A new approach to travel tackles congestion. News release, 10 August 2006. Sustrans, Bristol UK. Available www.sustrans.org.uk ⇨ Media ⇨ News releases.

## Chapter 5
# Climate Change Mitigation and Sustainability: Educational Issues

**Remigijus Ciegis, Dalia Streimikiene, Dalia Gineitiene**[1]

## Abstract

There is a dual relationship between climate change and sustainable development policies, thus climate change mitigation should be considered in line with policies aiming to implement sustainable development. Training, education and raising awareness are the crucial issues in seeking to implement climate change mitigation and sustainable development policies and especially now, when environmental regulation tools are becoming voluntary and ethical public participation is the most important issue in implementing policies. Only an educated society, which understands the importance of climate change mitigation and sustainable development principles, can create a favourable situation for the development of corporate social responsibility and other voluntary measures.

This chapter analyses the relationship between climate change and sustainable development and discusses the most important issues concerning education in this field. The analysis of the main problems and challenges of Parties to UNFCCC in climate change education is performed and comparison with Lithuania's situation in this field is provided. Based on analysis performed, the proposals for the development of education in Lithuania integrating climate change mitigation issues into the scope of sustainable development education were elaborated.

## Introduction

When addressing climate change mitigation and adaptation, it should be recognised that the directions of the development and the institutions involved in this as well as more specific mitigation and adaptation policy options influence them. This also means that policies influencing the development and institutions involved have indirect impacts on climate change adaptation and mitigation despite the fact that they are targeted towards broader development goals. These impacts can be positive or negative, and several studies have therefore suggested integration of climate change adaptation and mitigation perspectives into development policies in order to make development more sustainable (Davidson et al. 2003; Munasinghe 2002; Munashinge & Swart 2005).

---

1 Socio-Cultural Research Centre, Kaunas Faculty of Humanities, Vilnius University.

Arrow et al. (2004) presents an approach for evaluating alternative criteria for consumption over time seen from a sustainable development perspective. Intertemporal consumption and utility are introduced here as measurement points for sustainable development. One of the determinants of consumption and utility is the productive base of society, which consists of capital assets such as manufactured capital, human capital, and natural capital and institutions. Institutions can be understood as part of the capital assets as well as guiding the allocation of resources including capital assets. Institutions include the legal structure, formal and informal markets, various government agencies, interpersonal networks, and the rules and norms that guide their behaviour (Beg et al. 2002).

Seen from a sustainable development perspective, the issue is then, how and to which extent policies can influence the productive basis of society and thereby make development patterns more sustainable, which will imply the implementation of some climate change adaptation and mitigation policies.

There is a dual relationship between climate change and sustainable development policies, thus climate change mitigation should be considered in line for policies aiming to implement sustainable development. Training, education and raising awareness are the crucial issues in seeking to implement climate change mitigation and sustainable development policies. Especially today when environmental regulation tools are becoming voluntary and ethical public participation is the most important issue in implementing policies. Only an educated society understanding the importance of climate change mitigation and sustainable development can create a favourable situation for the development of corporate social responsibility and other voluntary measures and can force business to take responsibility for climate change mitigation and sustainable development as well as create situations where only environmentally and socially responsible business can gain a competitive advantage.

The *aim* of this chapter is to analyse the relationship between climate change and sustainable development and to raise the most important issues concerning education in this field.

The *main tasks* of the article are:

- to analyse the relationship between climate change and sustainable development;
- to analyse the main problems and challenges of parties to UNFCCC in climate change education, training and raising awareness based on their national communications to UNFCCC;
- to analyse the situation regarding climate change education, training and raising awareness in Lithuania based on III-IV national communication to UNFCCC;
- to overview education in Lithuania integrating climate change mitigation issues into the scope of sustainable development education.

## Climate change mitigation and sustainable development

The policy perspective can be the starting point in climate change adaptation and mitigation and sustainable development issues considered as an indirect policy impact. This approach will tend to focus only on more specific sectoral policies or climate policy instruments, which meet the adaptation and mitigation goals but are not necessarily supportive to all the economic, social, and environmental dimensions of sustainable development. This also implies that climate policy implementation can be hindered by tradeoffs between development goals and the global environment. Furthermore, climate policies that do not take economic and social considerations into account might not be sustainable in the long run. The following Figure 1 illustrates the interactions between the climate system, natural systems and society and the scope for policy interactions.

Fig. 1: Interactions between the climate system, nature, and the socio-economic system and the relationships to SD and adaptation-mitigation.[2]

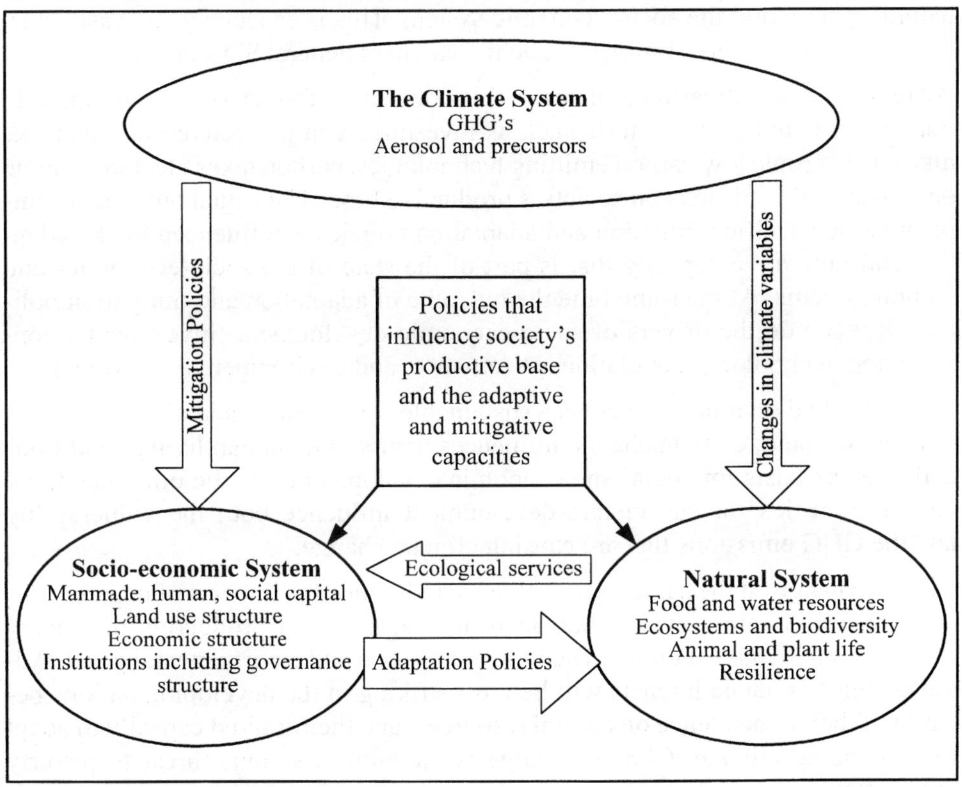


2 Source: IPCC. Third Assessment Report, Climate Change 2001: Mitigation.

Figure 1 includes three systems: the climate system, the natural system and the socio-economic system. The climate system is influenced by the GHG emissions, which result in climate change and that again puts a stress on the natural systems. The final impacts of climate change influence on the natural systems depend on the vulnerability and resilience of the system. As a result, the ecological services (such as resources for agriculture and fishery, aesthetic values or water resources) provided to the society are also influenced. Here it should be noted that all interactions between the systems have a complex and non-linear character (Atkinson et al. 1977).

The responses to climate change depend on society's productive base including capital assets and institutions as well as the adaptive and mitigative capacities. Policies that enhance the society's productive base and the adaptive and mitigative capacities are shown in Figure 1 in one box in order to emphasize many common features between these elements. Furthermore, policies in this area are supposed to have many synergetic impacts that at the same time influence both systems; the natural system and the socio-economic system. This is especially the case when the policies are targeted towards the achievement of general SD goals.

Adaptation and mitigation policies in this conceptual framework are assumed to be more isolated policy efforts such as investments in protective measures like dikes, new crops, low carbon emitting technologies, carbon taxes etc, that assume only marginal feedbacks on society's productive base. The actual outcome of implementing specific mitigation and adaptation policies is influenced by the adaptive and mitigative capacity that is part of the state of the socio-economic- and natural systems. At the same time the outcome of adaptation and mitigation policies depends on the drivers of socio-economic development (investments, consumption, technology, population, governance, and environmental priorities).

There is a dual relationship between sustainable development and climate change. On the one hand, climate change influences natural and human living conditions and thus the basis for social and economic development. On the other hand, society's priorities for sustainable development influence both the vulnerability and the GHG emissions that are causing climate change.

Climate change, at the same time, influences SD since the impacts of negative climate on ecosystem services, human health, agricultural production and many other areas will make it more difficult to meet social and environmental goals. Adverse climate change impacts will be most striking in the developing nations because of their dependence on natural resources, and their limited capacity to adapt to a changing climate. Climate change is therefore a serious threat to poverty eradication.

The Third Assessment of IPCC initiated a number of conceptual discussions about the relationship between SD and climate change. The Synthesis Report (IPCC 2002) recognises the importance of understanding the relationship between sustainable development and climate change and on this basis concludes that "the climate change issue is part of the larger challenge of sustainable development". This means that climate policies can be more effective when they are embedded within broader strategies designed to make national and regional development paths more sustainable. This is important because the ability of a country to achieve sustainable development goals is affected by the impact of climate variability and change, climate policy responses, and associated socio-economic development. Conversely, the pursuit of those goals will in turn affect the opportunities for, and success of, climate policies. In particular, the socio-economic and technological characteristics of different development paths strongly affect emissions, the rate and magnitude of climate change, climate change impacts, the capability to adapt, and the capacity to mitigate. The major findings of IPCC at the more conceptual level were the following: a) sustainable development can be used as a framework for understanding society's ability to respond to climate change impacts, but b) more work is needed to understand and assess the capacity for policy implementation.

The existing dual relationship between SD and climate change requires that education policies address both the SD and the climate change. Thus, the need for education, training and public awareness raising policies in order to pursue various projects and programmes on climate change is obvious and the target groups could be the following: a) pre-school children, pupils of secondary and other schools, b) students of universities and higher schools; c) local communities; d) business leaders; e) decision makers.

## Public awareness on climate change

National strategies addressing climate change could be successfully implemented only when the general public and important interest groups (stakeholders), such as businesses and local decision makers are involved. Businesses, communities, and individuals are directly responsible for most greenhouse gas emissions (GHGs). Governments must persuade them to adjust their activities in a way that reduces their direct emissions. Otherwise, it will not be possible to realize the objectives of the UN Convention on Climate Change and the targets of the Kyoto Protocol.

To achieve this goal governments have several tools at their disposal. They can use policies to raise the cost of activities that emit GHGs, regulations and standards mandating changes in products and practices as well as to adjust taxes and subsidies to influence behaviour. However, those concerned about the imposition of a new cost can oppose such policies and measures. Therefore, to gain public support for such policies the importance of public information and education is obvious. They can also encourage changes in public behaviour that will lead to lower emissions.

Many governments, intergovernmental and non-governmental organizations are already working actively to raise awareness. However, activities of a greater magnitude are needed because of the large scale of changes required and the vast number of people and interests to be reached. The UN Convention on Climate Change recognises this reality and Article 6 on Education, Training and Public Awareness requires governments to promote the development and implementation of educational and public awareness programmes, promote public access to information and public participation, and promote training of scientific, technical and managerial personnel. The Kyoto Protocol calls on parties to cooperate and to promote the development and implementation of educational and training programmes, including the strengthening of national capacity building, and to facilitate, at the national level, public awareness and public access to information.

The primary aim of public awareness campaigns is to enable the public to participate in activities addressing climate change because the involvement of the public depends in part on the level of their knowledge of this issue. In their national communications, countries not included in Annex I to the Convention indicated that their activities were targeted; to inform people about the causes and threats of climate change; to encourage actions through implementation of energy efficiency and other conservation measures; and to solicit feedback regarding national communications and national action plans. (UNFCCC 2005b). They used dissemination of printed materials, such as brochures, fliers, posters, pamphlets and technical publications and journals as the most common means to inform the public. The widespread availability of mass media; radio, print and television made it easier for countries to reach out to the general public on climate change issues.

In their national communications, countries included in Annex I to the Convention reported their activities and noted the importance of public participation and public access to information. Regarding the access to information, countries stressed the increasing opportunities in relation to the emerging use of electronic tools, which, in addition, enhance partnerships and networking in this area. They recognized that in order to increase public understanding of climate change and its impacts, actions to reduce emissions and a need for long-term sustained efforts is obvious. (UNFCCC 2003) Collaboration with non-governmental organizations as well as with private sector is important in training and informing the public and gaining support for policies and measures related to climate change mitigation.

## Educational issues of climate change mitigation

Education, training and public awareness are important in implementing climate change projects and programmes. Although the specific objectives and priorities for public education differ according to the needs of the target sector, the overall goal is still to ensure that every person understands the problems associated with climate change and assumes some responsibility to address these problems:

- to educate the public on the causes and effects of climate change;
- to strengthen institutions involved in project implementation and coordination;
- to mobilize the public to implement projects, e.g. renewable energy, energy efficiency and forest and coastal resource management;
- to train educators to develop and integrate climate change into the school curricula at primary, secondary and tertiary levels;
- to raise the level of awareness of policy and decision makers on climate change;
- to build the skills and knowledge of local experts;
- to disseminate information on national communications;
- to strengthen national capacity to develop and implement climate change action plans. (UNFCCC 2005)

All parties of UNFCCC in their national communications are required to report on activities in education, training and public raising awareness in the climate change field. Sixth Compilation and Synthesis of Initial National Communications from Parties not Included in Annex I to the UNFCCC elaborated in 2005 summarized information from parties of UNFCCC on education, training and public awareness. Lithuania faces similar problems to other transition and developing economies, therefore the problems of developing countries in this field are discussed further. Based on the analysis provided by Sixth Compilation and Synthesis it is seen that the most important problem related to climate change is that it is a relatively new area of study and most countries expressed the need for assistance in incorporating climate change into regular school curricula. Parties of UNFCCC also identified the need for support in strengthening academic and research institutions to serve as centres of excellence on climate change by educating students who are interested in climate change issues.

Since climate change affects the lives of every individual, countries targeted their activities towards the following stakeholders: children, youth, students, teachers, researchers, policy makers, parliamentarians, local government officials, non-governmental organizations, community-based organizations, media, the private sector and industry. Some countries emphasized that more focused education campaigns aimed at specific or vulnerable stakeholders, e.g. women, farmers, etc., were needed. Education, training and public awareness activities on climate change are undertaken in cooperation with different agencies and organizations.

The Internet allows for greater transparency, increased coordination among agencies involved in climate change issues and heightened public awareness, thus countries disseminated information through the Internet and several countries set up web pages on climate change that include information on the preparation of their initial national communications and updates about climate change. Internet communication also provides an opportunity for national experts and the public to provide feedback.

The need for education, training and public awareness in order to pursue various projects and programmes on climate change was emphasized. The difficulties encountered in implementing education, training and public awareness projects due to limited financial and human resources were acknowledged.

The shortage of experts and professionals working on climate change is seen as a barrier to implementing projects relating to climate change. Since the availability of experts is very important for the implementation of climate change projects, the establishment of continuous training programmes should ensure it. Education plays an important role in ensuring the development of a critical mass of experts on climate change. Most countries reported that governmental efforts and initiatives from the academic sector are under way to mainstream climate change into the different school curricula and to address the lack of education on this issue. In this respect there is the need for support in developing professional training packages on GHG inventories, education, training and public awareness, vulnerability and adaptation assessments, mitigation assessments and technology needs assessments.

In spite of these challenges, countries pursue education, training and public awareness activities at the national, sub-regional and regional levels.

## Integration of climate change into basic education

According to the data of the Sixth Compilation and Synthesis (UNFCCC 2005b) some countries have developed national policies ensuring the continuity of environmental education programmes and aimed at integrating environmental issues and climate change into basic subjects in education, for example, the National Environmental Education Policy (Brazil), Guidance on Environmental Education in Primary and Secondary Education (China), and the National Strategy on Environmental Education (Cuba).

Although most countries are facing problems integrating climate change into different levels of education, progress is reported in this area. Materials on climate change are prepared and disseminated to primary, high, and technical schools in many cases as the result of collaboration within different projects. As an example Pakistan could be mentioned, the country with support from the Swiss Government, plans to integrate environmental issues, including climate change, into relevant textbooks through the Environmental Education Promotion at School and College Level project.

Among the problems related to the implementation of educational programmes, countries mentioned lack of instructional materials as well as teachers qualified to instruct on climate change issues as barriers. Thus, countries plan to develop materials on climate change and to train teachers. For example, the Ministry of Education in Kenya, in cooperation with various agencies, plans to develop instructional modules on climate change.

Thus, regarding basic schools, in order to integrate climate change issues into environmental education programmes it is important to develop national policies integrating environmental issues, preparing relevant materials and textbooks and training teachers.

## Integration of climate change into university courses

As seen from the Sixth Compilation and Synthesis (UNFCCC 2005b) in some countries climate change is incorporated into university courses, such as geography, forestry and natural resource management, ecology, biology, physics and meteorology. In other countries they have established university courses on the environment and revised existing curricula on environment, energy and atmospheric studies to include elements of climate change. A number of universities offer undergraduate and post-graduate programmes and research opportunities on climate change.

Few countries indicated that the scholarships from governments and/or bilateral and multilateral organizations are offered to undertake short- or long-term courses at national or foreign universities. The recipients of these scholarship programmes include students, researchers, media practitioners and civil servants working in environmental agencies. For example, the National Council for Scientific and Technological Development of Brazil offered 20 scholarships for doctoral and post-doctoral studies; South Africa is funding research on climate change through its National Research Foundation.

Parties not included in Annex I to UNFCCC reported receiving training in connection with the preparations for the initial national communications and implementation of various projects involving renewable energy and energy efficiency, vulnerability and adaptation assessments. These programmes were aimed at building the capacities of national experts and staff involved in operating institutions and government agencies involved in climate change. In some countries training of staff was organized into subjects such as greenhouse gas (GHG) inventory preparation, vulnerability and adaptation assessment, and identification of mitigation measures. Training assistance to prepare their initial national communications was received through the GEF and through various projects funded by bilateral and multilateral institutions. Also workshops and seminars were organized relating to the preparation of national communications, specifically on emission factors and activity data, mitigation analysis, adaptation options, predictive computer modelling, technology needs assessments and Intergovernmental Panel on Climate Change methodology.

Specialized training and seminars were provided for government officials and legislators to facilitate the development and implementation of policies relating to climate change. In some countries environment and climate change was integrated into the training courses offered for senior and mid-level government officials through their national schools of public administration. The participation of the experts and senior officials in the relevant international events was seen as a means of increasing their knowledge about the Convention and enhancing their skills in negotiations. It should be also mentioned that countries stressed, on the one hand, the very important role of local governments in the development and implementation of policies and measures to address climate change but, on the other, the lack of capacity in this field and the need to establish training programmes for local government officials is obvious.

One of the problems related to training experts and educating and raising the awareness of the public in most countries is limited financial and human resources and competing priorities, such as poverty alleviation and food security. The lack of experts working on climate change issues is partly due to the high turnover of experts involved in climate change. Thus, countries see the need to strengthen governmental, non-governmental and academic institutions in order to sustain the programmes aimed at building and enhancing local capacities to respond to climate change.

## Climate change integration into education for sustainable development

Only an educated society that understands the importance of climate change mitigation and implementation of sustainable development principles can create a favourable situation for the development of corporate social responsibility for climate change mitigation. The dual relationship between sustainable development and climate change points to the need for education policies that address both sustainable development and climate change. Thus, climate change mitigation issues could be integrated into the education for sustainable development (ESD).

The documents for the UN Decade of ESD distinguishes four major domains of ESD, reflecting diverse goals and audiences: promotion and improvement of basic education, reorienting existing education at all levels to address sustainable development, developing public understanding and awareness of sustainability, and training. Within these domains the following issues are indicated as the priorities: overcoming poverty, gender equality, health promotion, environmental conservation and protection, rural transformation, human rights, intercultural understanding and peace, sustainable production and consumption, information and communication technologies, cultural diversity (UNESCO 2003). Because of the integrative approach of ESD it encompasses various interrelated environmental, economic and social issues, thus, ESD can be used as a framework for increasing society's understanding and ability to respond to climate change impacts. It must be seen as part of a lifelong learning process taking into account the changing natural conditions associated with economic, social and cultural development.

The Strategy for Education for Sustainable Development (UNECE 2005) points out that in order for ESD to become part of the lifelong learning of individuals including the public administration sector, the private sector, industry, transport and agriculture, it is necessary to: "offer SD-related learning opportunities in continuing education for professionals, including those in planning, management and the mass media; encourage and support community-based SD-awareness-raising activities; develop cooperation with NGOs and support their educational activities; promote cooperation among formal educational institutions and non-formal organizations as well as informal activities; encourage the media to inform and debate issues for sustainable development to reach the general public".

Although formal education in schools and universities has the main role due to the concentration of knowledge and skills, other forms of education are also very important in relation to educating a participating citizen. ESD could be seen as a cross-sectoral approach covering the main forms of learning and education:

- *formal* learning (leading to recognised diplomas and qualifications such as secondary, higher education);
- *non-formal* learning (not leading to formal certificates and provided at the workplace and through the activities of civil society, organizations and groups, e.g. NGOs, nature centres, etc.)
- *informal* learning (everyday life and non-intentional learning contributing to their knowledge and skills, such as, among others, the media).

Among the challenges it is possible to mention the need for integration of climate change mitigation issues into the educational programmes and strengthening coordination and collaboration between different levels of education – formal, non-formal and informal learning. The Strategy for Education for Sustainable Development stresses the importance of formal education institutions in developing capacities from an early age, providing knowledge and influencing attitudes and behaviour while non-formal education is often more learner-oriented, participatory and promotes lifelong learning. Thus, "non-formal and informal learning, including public awareness programmes, should aim to provide a better understanding of the links between social, economic and environmental issues in local and global contexts, including a time perspective" (UNECE 2005).

The content of education and learning must embrace interdisciplinary, system thinking to address environmentally sustainable action on local, regional and global scales over short, medium and inter-generational time periods. An important factor for promoting and facilitating the work is good educational and informational materials at all levels. This requires access to different information resources including textbooks, cases studies and good practices as well as media, web-resources, etc. There is a problem that not all educators are familiar with the sustainable development approach and thus cannot incorporate it or climate change issues into their teaching. Problems of understanding also exist among decision makers at different levels, in the business community as well as among the public.

Thus, ESD must be seen as part of a lifelong learning process taking into account the changing ecological conditions associated with economic, social and cultural development. ESD enables all individuals to fully develop the knowledge, perspectives, values and skills necessary to take part in decisions to improve quality of life both locally and globally in terms that are most relevant to their daily lives.

## The Situation in Lithuania

Lithuania reported in National Communication to UNFCCC (2005a) that the country has appropriate legislation for the development of a modern educational system and social partnership. The National Strategy for Sustainable Development (2003) also foresees that the sector of global climate change is integrated into long-term goals of education and science. Among the goals foreseen in the Strategy are to build a society of educated, self-dependent, active and responsible members, to develop its intellectual potential and thus ensure the coherent and harmonious implementation of economic, social and environmental aims.

At present the existing legal system in Lithuania guarantees its citizens the right of access to environmental information and participation in decision-making. These rights are ensured by the Aarhus convention ratified by Lithuania, corresponding EU Directives as well as the laws of the Republic of Lithuania, resolutions of the government and ministries, orders and ordinances. The Government Resolution, “On the Procedure for the Provision of Information on the Environment to the Public” approved in 1999, is one of the main documents. Its objective is to guarantee the right of access to environmental information held by or for public authorities and to set out the basic terms and conditions of, and the practical arrangements for, its exercise. The Ministry of Environment should be mentioned as one of the main institutions responsible for the accessibility of environmental information and implementation laws as well as other documents. The Public Information and Public Relations Department, among its other functions is responsible for framing and pursuing the policy of public environmental education and information, ensuring provision of environmental information and its dissemination to the public, promoting activities of state institutions and NGOs as well as enhancing their cooperation in public environmental education, and coordinating projects on environmental education (UNFCCC 2005a). The Ministry of Environment has special web pages designed for UNFCCC issues.

Issues of public environmental education and information about climate change-related problems and ways to address them are included in the National Strategy and Action Programme for Environmental Education, Information and Public Awareness of Society, prepared by the Ministry of Environment in 1998.

Education for sustainable development is promoted and recognized as important to the success of sustainable development and the process of reorientation of education to address sustainable development. It is especially felt after the UNECE Strategy for Education for Sustainable Development was adopted at the high-level meeting of Environment and Education Ministries in Vilnius in 2005.

The system of formal education including universities, high schools and secondary schools has sustainable development as well as global climate change issues in their long-term goals of education and science.

Study programmes offered by higher educational institutions of Lithuania (Vilnius University, Vilnius Pedagogical University, Klaipėda University, Šiauliai University, Vilnius Gediminas Technical University, Vytautas Magnus University, Kaunas University of Technology, and the Lithuanian University of Agriculture) cover different topics on sustainable development. Such topics as vulnerability and sensitivity of ecosystems, global climate change impacts on economy, ecosystems and society, adaptation, mitigation measures are among the climate change and related topics (UNFCCC 2005a).

The results of the research projects implemented by higher institutions were presented and problems of climate change trends and impacts were raised at various different levels; international, regional and national conferences organized by the Ministry of Environment, Vilnius University, the Institute of Ecology of Vilnius University, Lithuanian Academy of Sciences, Klaipėda University, Vilnius Gediminas Technical University, and others.

However, universities, which make a large economic entity consuming a big amount of resources purchasing products and services, still do not discuss how to reduce the environmental impact of their own operations, and also do not tie these efforts to the formal curriculum.

Requirements for the quality of basic and secondary education are set in the "General Programs and Education Standards"[3]. Environmental education is integrated and different aspects of various environmental issues including climate change are discussed when teaching separate subjects (biology, chemistry, physics, etc.), thus teachers of different specialties have to coordinate their efforts. Education for sustainable development and principles of sustainable development are introduced in schools into the educational content and process during implementation of "Agenda 21" related projects. Ministry of Science and Educa-

---

3 www.smm.lt.

tion has prepared a project on integrated programmes that foresees the inclusion of sustainable development issues into the content of different programmes.

Non-formal education in Lithuania covers various governmental and non-governmental institutions that take part in environmental education and the raising of public awareness. A significant role in the implementation of the Convention and information and education of the public is performed by the environmental NGOs. At present 80 environmental NGOs form a coalition, which was established in 2005. They organize different educational events, produce leaflets, brochures, books, and disseminate information using mass media and Internet sites. For example, the NGO Lithuanian Ecological Movement regularly publishes articles on matters related to climate change. In 2006, as a result of a project "Awareness Raising Campaign on Climate Change", which was co-ordinated by the NGO Environmental Centre for Administration and Technology and financed by the Embassy of the United Kingdom in Lithuania, recommendations for Education on Climate Change were prepared. Unfortunately, due to limited human and financial resources, the activities of NGOs do not always reach a large number of the population.

It is noted in the national communication of Lithuania that in many cases the main drivers behind activities in the field of education and raising public awareness in Lithuania are highly motivated environmental officials, experts, lecturers and teachers who try to include climate change and environmental issues in their daily work. The low level of activities and number of people targeted resulted in a limited awareness of climate change among the general public and some governmental institutions. However, it should be noted that in recent years public attention to climate change in Lithuania is continuously increasing. This is partially related to the activities of environmental NGOs, local governments but mostly due to special publications in the press, as well as TV and radio reports and interviews on climate change issues made by Lithuanian climate and environment protection specialists.

## Conclusions

1. There is a dual relationship between sustainable development and climate change. On the one hand, climate change influences natural and human living conditions and thus is the basis for social and economic development. On the other hand, society's priorities for sustainable development influence both the vulnerability and the GHG emissions that are causing climate change. This existing dual relationship between SD and climate change requires education policies that address both the SD and the climate change.
2. National strategies addressing climate change could be successfully implemented only when the general public and important interest groups ("stake-

holders"), such as businesses and local decision makers are involved. The primary aim of public awareness campaigns is to enable the public to participate in activities addressing climate change because the involvement of the public depends in part on the level of their knowledge of this issue.

3. Based on the analysis provided by Sixth Compilation and Synthesis it is seen that the most important problem related to climate change is that it is a relatively new area of study and most countries expressed the need for assistance in incorporating climate change into regular school curricula.
4. Regarding basic schools, in order to integrate climate change issues into environmental education programmes, it is important to develop national policies integrating environmental issues, preparing relevant materials and textbooks and training teachers.
5. Countries stressed, on the one hand, the very important role of local governments in the development and implementation of policies and measures to address climate change but, on the other hand, the lack of capacity in this field and the need to establish training programmes for local government officials is obvious.
6. Because of the integrative approach of ESD, it encompasses various interrelated environmental, economic and social issues, thus ESD can be used as a framework for increasing society's understanding and ability to respond to climate change impacts.
7. The content of education and learning must embrace interdisciplinary, system thinking to address environmentally sustainable action on local, regional and global scales over short, medium and inter-generational time periods.
8. In Lithuania, education for sustainable development is promoted and recognized as important to the success of sustainable development as is the process of reorientation of education to address sustainable development.

## References

*Arrow, K., Dasgupta, P., Goulder, L., Daily, G., Ehrlich, P., Heal, G., Levin, S., Mäler, K.G., Schneider, S., Starret, D. & Walker, B.* (2004), „Are We Consuming Too Much?". Journal of Economic Perspectives. vol. 18, no. 3, pp. 147–172.

*Atkinson, G., Dubourg, R., Hamilton, K., Monasinghe, M., Pearce, D &, Young, C.* (1997), Measuring Sustainable Development. Macroeconomics and the Environment, Edward Elgar, London.

*Beg, N., Morlot, J.C., Davidson, O., Afrane-Okesse, Y., Tyani, L., Denton, F., Sokona, Y., Thomas, J.P., Rovere, E.L., Parikh, J., Parikh, K. & Rahman, A.* (2002), „Linkages between Climate Change and 4 Sustainable Development". Climate Policy, vol. 2, pp. 129–144.

*Davidson, K., Halsnaes, S., Huq, M., Kok, B., Metz, Y., Sokona, J. & Verhagen.* (2003), „Development and climate: the case of Sub-Saharan Africa", Climate Policy, vol. 3 (suppl. 1), pp. 97–113.

*Munasinghe, M.* (2002), „The sustainomics trans-disciplinary meta-framework for making development more sustainable: applications to energy issues". International Journal of Sustainable Development, vol. 4, no. 2, pp. 6–54.

*Munasinghe, M. & Swart, R.* (2005), Primer on Climate Change and Sustainable Development. Cambridge University Press, UK.

IPCC (2002), Third Assessment Report, Climate Change 2001: Mitigation. The Edinburgh Building Shaftesbury Road, Cambridge, England.

NFCCC (2005a), Lithuania's Third and Fourth National Communication on Climate Change. Ministry of Environment of Republic of Lithuania, Institute of Ecology of Vilnius University, Vilnius.

UNECE (2005). UNECE Strategy for Education for Sustainable Development, Geneva.

UNESCO (2003). Framework for a Draft International Implementation Scheme. Accessed (December 2005) www.cee.org.uk.

UNFCCC (2003), National communications from Parties included in Annex I to the Convention. Compilation and synthesis of third national communications. Accessed (December 2006) http:// unfccc.int/ resource/ docs/ 2003/ sbi/ 07.pdf.

– (2005b), National communications from Parties. Sixth compilation and synthesis of initial national communications from Parties not included in Annex I to the Convention. Accessed (December 2006) http://unfccc.int/ resource/ docs/ 2005/ eng/ 18a06.pdf.

The Lithuanian Strategy for Sustainable Development (2003), Vilnius: Lutute.

## Chapter 6

# Case Study Research as Bridge Builder between Science and the Society – The Rationale behind the ASTRA 2005 Winter Storm Study

**Simo Haanpää, Lasse Peltonen**

## Abstract

The challenge of climate change as a complex global issue is that it demands locally grounded solutions that bring together actors from various fields. As a 'wicked' problem climate change adaptation demands deliberation between different subsystems of society, including scientific research. While answering the complex needs of the society, it has to be re-evaluated what 'good' science means. This brings about a need for socially robust knowledge.

The need for contextualisation of scientific knowledge is supported by theoretical development in the fields of policy analysis and risk governance. It also entails an epistemic shift away from the 'quest for certainty' towards a pragmatist understanding of knowledge. The underlying request is to integrate expert knowledge with the needs of other stakeholders – to bring in the people to add to the policy formation a notion of emotional knowledge.

As a research method, case studies readily contextualise scientific information and therefore offer valuable insights into the underlying social values of the problems at hand. As locally grounded narratives they can offer a shortcut to the formation of a new kind of expertise needed as the demands from the society reform the claims on scientific information.

It is argued that developing the science-policy interface and risk communication can both benefit from a contextual research approach using case studies. Such studies have critical and persuasive functions, and they provide a basis for further case-based learning exercises with opportunities for awareness raising, institutional capacity building and practical adaptation measures.

## Introduction

A severe winter storm developed in the Northern Atlantic in January 2005. The storm, 'Gudrun', first hit the British Isles and then continued towards the Baltic Sea. The storm was the fiercest to hit the Baltic Sea since 1969; it left 17 casualties and caused extensive material damage, estimated at a total of EUR 2,4 billion (with insured losses exceeding EUR 1,5 billion). Hundreds of thousands of people were affected by the coastal storm surges, floods, electric blackouts and transport cancellations caused by the storm.

Such an extreme event captures the attention of the media and the public in Europe far more efficiently than any debate on „global environmental change". While the storm presents a local and immediate threat, climate change is often seen as global and long-term issue whose effects are mostly felt by far away developing countries. This discrepancy between the abstract and distant perceptions of global climate change and the very concrete effects of the 2005 winter storm were the starting point for a case study – and a risk communication effort in the Interreg IIIB project ASTRA (Developing Policies and Adaptation for Climate Change adaptation in the Baltic Sea Region). The case study was used to further the main goal of the project, namely, to raise awareness of and develop response strategies to climate change adaptation.[1]

In the following, the issue and challenges of communication between science (or, more broadly, knowledge production) and politics (or, policymaking and decision-making) is discussed in the context of climate change adaptation. We are especially interested in the role of case studies in mediating scientific knowledge and guiding the climate change adaptation process. We argue that developing the science-policy interface and risk communication can both benefit from a contextual research through case studies in the field of climate change adaptation. Such studies provide a basis for further case-based learning exercise which provides opportunities for awareness raising, institutional capacity building and practical adaptation measures.

From a policy perspective, we argue that policies chosen can never be based on scientific data alone, but that they contain considerable amount of practical, or rather emotional, knowledge, as Fischer (2006)[2] puts it. Therefore, our interest in terms of case studies lies in their role as a basis for local narratives and through that in their power to change the practical discourse affecting the actions of actors. In practical terms, in the centre of the problem of adapting to climate change lies the question of how to balance between the need for top-down national policies and locally grounded ways to implement them.

---

1 ASTRA is funded by the Interreg IIIB Baltic Sea Region (BSR) starting from 2005 until the end of 2007. The project has brought together scientists working in the fields of human and natural sciences, and various stakeholders from cities and regions in the BSR, including NGOs. The project deals with a wide scope of climate change effects studied in the context of the Baltic Sea region. Disseminating the results to planners is a key interest. See more at: www.astra-project.org/cms/

2 Fischer (2006), Demeritt (2006), Throgmorton (2006) and Peuhkuri (2006) refer to presentations and comments given at the YHYS Conference 2006, themed "Case studies in environmental social science". The conference was held at the University of Tampere, Finland, on the 22nd -23rd of November, 2006.

In this chapter, we will look at the theoretical grounds for a more contextual approach to climate change research. First introducing topical theoretical developments in the fields of policy formation and risk communication, we turn the discussion on practical level through introducing a case study on winter storms made under ASTRA-project[3].

## The epistemological perspective: certainty versus pragmatism

While case studies have been used in hazard research, where practical improvements in safety measures are of essence, the case study method has not been part of climate change research. One reason for this lies in the fact that climate change research has focused on producing certainty about the phenomenon itself. Research efforts have been concentrated on proving that the climate is changing and that this change can be attributed to human action. This 'quest for certainty' (Dewey 1929) has led to the dominance of natural sciences and a limited role for social sciences and qualitative research. Interestingly, David Demeritt (2001) argues that the science of climate change entails epistemic commitments that support a technocratic approach to managing the problem. According to Demeritt, the intertwining of science and politics has produced a dominant understanding of climate change as defined by natural science and climate modelling. Climate change is primarily perceived as a global-scale phenomenon, which has sidetracked more reflexive accounts of climate change from other perspectives of knowledge production and spatial scale, including the local.

In any case, the problems of climate change adaptation provide a challenging context for policy analysis and risk communication. This is partly due to the complexity of climate change itself, but also due to the complexity of communications between science and society and – following Demeritt's argument above – due to a mismatch between the dominant understanding of science and practical needs of society. These issues become increasingly important in any effort to advance climate change adaptation.

First, there is the complexity of climate change itself, as events related to climate change are highly uncertain, and the outcomes poorly defined. Both climate change and the societies with which it interacts are highly complex and dynamic entities. The kinds of uncertainties related to climate change include epistemic uncertainty (related to incomplete knowledge), natural stochastic uncertainty (relating to the chaotic nature of the climate system) and human reflexive uncertainty (relating to human agency in terms of greenhouse gas emissions, mitigation and adaptation measures) (Patt & Dessai 2005).

---

3 Haanpää et al. (2006) Impacts of winter storm Gudrun of 7th-9th January 2005 and measures taken in Baltic Sea Region. Unpublished working paper for ASTRA-project, to be downloaded at www.astra-project.org.

Second, the communication between the fields of science, planning and decision-making is far from straightforward. The two groups have their personal interests and codes of professional conduct, which affect their actions. Referring to the relation between scientists and so-called „end-users", Nowotny (2000) asks:

> „What are the conceptions that engineers, climate modellers, molecular biologists and others have about people – as individuals and collectives – who will utilise or be affected by the knowledge they produce?"

Nowotny argues that such conceptions are usually not very well articulated. She also argues that science is increasingly being critically reviewed and the end-users of scientific knowledge are more and more educated and reflexive towards the „products" of science. This means that „society talks back to science" (Nowotny et al. 2001) In sum, the context of scientific knowledge influences how its reliability and objectivity are assessed. Nowotny (2000) argues that contextualisation can make knowledge robust and operational. Climate scenarios, for instance, are not universally reliable and useful in planning for adaptation measures. Dessai et al. (2005) have pointed out that the role played by climate scenarios depends on many contextual factors such as the chosen adaptation assessment approach and the availability of technical and financial capacity to handle scenario information.

Due to complex contextuality, any social response to climate change (mitigation or adaptation) can be seen as a 'wicked problem', a term coined by Rittel and Webber (1973) in their critique of the idea of synoptic, science-driven planning. According to Rittel and Webber, such 'wicked problems' have several problematic characteristics: First, they are too multifaceted to be definitively described. Second, in a pluralistic society there is nothing like the undisputable public good, and there is no objective definition of equity. Third, policies that respond to the problem cannot be simply correct or false, only better or worse – and this depends on multiple criteria. Finally, it makes no sense to talk about optimal solutions to climate change problems unless severe qualifications are imposed first. Even worse, there are no solutions in the sense of definitive and objective answers.

Developing any kind of climate change adaptation measures or policy, then, cannot be a purely scientific endeavour. To unfold the discussion behind the policy cycle formation we must start by investigating how the joint approval of the 'problem' is formed. This requires an understanding of the actors' needs, along with their situated knowledge and specific constraints. As we have seen, problem framing is never based on scientific data alone.

Demeritt and Langdon (2004) even challenge the often stated view, that the actors are lacking accurate scientific information on the regional impacts of climate change. While studying the viewpoints of local actors in Britain, they found no correlation between perceived accuracy and appropriateness to the local actors' needs. This led them to believe that this sort of technical problems referred to in open questions of their study could rather indicate the troubles that local actors have with coping with 'limited resources, competing priorities and little responsibility for key areas of mitigation and adaptation' (Demeritt & Langdon 2004). Although part of the problem framing has to do with multiple subsystems interacting, there is thus a practical component, too, that hinders the abilities of planners to take full use of the scientific information available: time.

This came up in a study on the flood risk in Finland[4], where the interviewed planners stated that both their resources to follow the scientific discussion on climate change and the possibilities to implement planning decisions solely influenced by the foreseen effects of climate change are severely congested by other factors of spatial development (Peltonen et al. 2006). Therefore, although scientific agencies are seen as the most trustworthy sources in many studies (like in those made in the UK on planner's perceptions on climate change (Demeritt & Langdon 2004) and on the level of public's knowledge on climate change in Finland (Eboreime & Wennerström 2005), the results throw a cast on the frequency on which these sources are consulted.

Demeritt and Langdon (2004) end up looking at the issue from an another angle: instead of being accurate, the information given to actors has to be trustworthy and easy to understand – and not only for the planner, but also for those the broader public. This theme resonates with pragmatist thought: the quest for certainty can, in fact, be at odds with practical concerns and learning. A related issue is the misunderstanding about what it actually means to disseminate scientific knowledge. According to the UK study, media sources of climate change were used almost as often as the official ones (Demeritt & Langdon 2004). Demeritt and Langdon (2004) explain this quickly by the easiness of understanding media sources and conclude, that the problem of disseminating climate change information is rather than technical or informational, more political and organisational (Demeritt & Langdon 2004).

## The perspective of policy analysis – from policy advice to contextual narrative

Policy analysis (or, policy science), is a field of research which is specifically interested in the production and use of knowledge in decision-making. The basic goal of policy analysis is to *provide political decision-makers with sound and relevant information and knowledge to enable collective decision-making in democratic, political settings*. From this perspective, the product of policy analysis

4 EXTREFLOOD (Tulvariskien hallinta yhdyskuntasuunnittelussa – Managing flood risk in spatial planning). Financed by the Ministry of Environment and Ministry of Agriculture and Forestry, 2004–2005 (See Peltonen et al. 2006)

is good *advice* (Weimer & Vining 1992)[5]. Policy advice, however, does not take place in a vacuum. Policy formation has been defined as a 'web of decisions and actions that allocate ... values' (Easton 1953, cit. Adger et al. 2004). Policies are, consequently, channels for values of the society, which are thus embedded in the very core of the policies themselves that guide the actions of actors. As Weimer and Vining (1992) note, good policy analysis takes a comprehensive view of consequences and social values. Social values come into play even where policy analysis is seen as purely instrumental and predictive.

Recently, policy analysis has countered an „argumentative turn", where policy scholars have been increasingly interested in the argumentative, socially constructed nature of policy and planning processes (Fischer & Forester 1993). This in turn has been coupled with a critique of positivist stances in planning and policy-making. Critical planning and policy scholars remind us that the context of arguments and discourse matters, and the staging and dramaturgy of policy deliberations are important factors in the construction of planning and policy-making. (Hajer 2001, 2002). This implies that communication between scientists and politicians *is not a matter of individual-level communications and simple advice*. Everybody can agree that policy research should provide decision-makers with sound and relevant information. However, the *interpretation* is often the problem: what actually constitutes „sound and relevant information and knowledge"?

The so-called postpositivist conception of policy science has developed a critique of the positivist mode of policy science, which sought to provide predictive causal analysis to enable optimal interventions by policymakers. Instead, the postpositivist conception of policy science addresses „the multidimensional complexity of social reality", and takes a discursive approach situating the empirical inquiry in a broader interpretive framework (p. 129). As such, decision-making is no longer seen through the narrow lense of the isolate, individual decision-maker acting upon objective knowledge in the optimal way (Fischer 1998).

Fischer (2006) talks about emotional knowledge, which includes the social values, cultural rationality and empirical and historical experiences of the public. These are formed and refined through the practical discourse in which we daily engage ourselves, including the impact of the media (Kunelius 2003). This context determines how we deal with the perception of risks. In a similar vein, Throgmorton (2006) argues that meanings are shaped through persuasive argumentation, the persuasiveness of which depends on the context. Therefore, it should be seen that what we are dealing with when we study local adaptive capacities, is *a web of locally grounded narratives* that form the local *expertise* that adaptive actions depend on.

---

5 However, policy analysis can also take a distance from this "servant" role. From this perspective, policy analysis creates a better understanding of policy processes. Thus, policy analysis has a dual nature, producing knowledge both *for the purposes of* decision-making and, perhaps more critically, *about* decision-making.

Fischer (2006) states that there is „an informal logic to everyday discourse". People generally do not act rationally in the positivist sense, but rather through everyday reasoning, rational with their emotional knowledge, including among other things their fears. Decision making is a cognitive process, where every situation is dealt with through past experiences. This is why factors such as (un)voluntarity, (un)familiarity, immediateness, (in)visibility and (un)fairness of an incident can actually override scientific data available. (Fischer 2006; Flyvbjerg 2001.)

In sum, policy scholars have developed an understanding of what policy advice may mean from a contextual perspective. The postpositivist perspective also holds for policy evaluation, at the other end of the policy cycle. Arguing for a better understanding of institutional and other constraints in policy implementation and evaluation, Majone (1989, 118) notes that „a practical problem is not solved by offering a theoretical solution that does not take into consideration the limitations which the context imposes". This also holds for the evaluation of different policy instruments: since the instruments are always used in some specific context, their evaluation should also be contextual, i.e. based on the constraints and opportunities of the setting where these instruments are to be applied.

This line of thought is carried further by Fischer (2006) who argues that we live in a 'world of action', where universal technologies cast into our societies are processed and adapted for practical purposes by the local and national societal context on which it is implemented. Fischer (1995; 2006) further proposes that policy evaluation should be widened from the pure technical verification of the policies to cover the justification of norms and standards underlying the scope of evaluation. The idea is presented in fig. 1 (according to Fischer 1995):

Fig. 1: The logical structure of a comprehensive evaluation

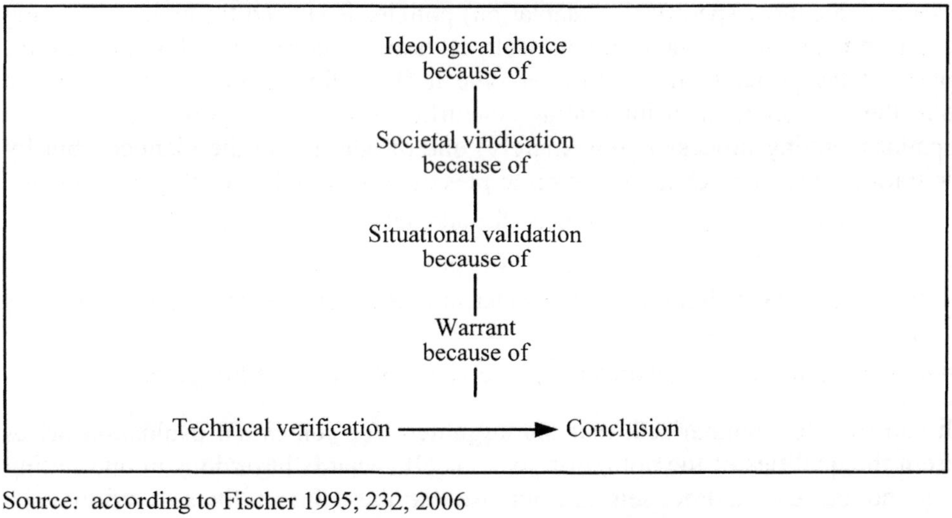


Source: according to Fischer 1995; 232, 2006

Here, the first-order evaluation covers the technical verification and situational validation that focus on the outcomes and objectives of the policies. These are the practical questions that determine the success of the policy in national and regional context. Indeed, on a practical level, as scientists often stress the uncertainties the climate models, emission scenarios and the interpretations of these provide, different actors generally want weighted information or rather certain 'forecasts' on how the phenomena are going to evolve in the future. Nowotny (2000), Dessai et al. (2005) and Amendola (2002) evaluate the role played by scientific data in terms of contextualising data through downscaling and revising its appropriateness in terms of the chosen adaptation measures. In previous studies this contextualisation has been made through downscaling climate models for the regional scale and by using case specific research in determining the most important regional issues. To determine local priorities for planning, mainly planners have been heard as actors (Lehtonen & Peltonen 2006).

This approach, as elaborate it is, gives less attention to one important feature mentioned by Amnedola (2002) however, that are the values and perceptions of the community. For the societal vindication and ideological choice behind policy formation we therefore have to look at the policies in wider context. These second-order levels look at the goals and values of the policies, in other words their instrumental or contributive value for the society as a whole and the legitimacy of the ideology behind the scope of the policies. This once again brings all stakeholders, including the public, to take part in the discussion on the policies, as the norms and standards affecting behind the warrant must be jointly approved (Fischer 2006).

Closing the gap in the policy cycle between policy formation and policy evaluation, Adger et al. (2004) argue that the preferred course of (in his study environmental, here more specifically adaptation) policies is shaped through an ongoing chain of decisions on the problem itself, on how to address it and on how to implement the policies chosen to overcome it. Therefore, it may well be argued that the real problem in integrating scientific results in the policy making and spatial planning process is not (merely) one of addressing the planners, but by initiating a broader debate on the principles of climate policy evaluation so that

1. the ideological choices of the need for adaptation are seen justifiable
2. the societal vindication of adaptive actions is accepted
3. the situational validation of the most important effects of climate change is accepted
4. a warrant for action towards the perceived problem is readily given

It can thus be summarised, that one argument for contextual evaluation arises from the qualities of the notion of „risk" itself (what is hazardous in one setting may not be so in another setting), but also that the vulnerabilities of individuals, households, organisations and cities depend on the particular historical paths that constitute these entities. Nothing is hazardous as such, since the whole notion of hazard and risk is relational – it begs for the question: *hazardous to what?*, *risky for whom?* Thirdly, as Renn (2001) notes, the perception of risk has contextual attributes and is perceived by far more numerous factors than simply by the probability and magnitude of harm. Renn goes on to list a number of factors affecting on our perception on risk that echo the ideas of Fischer presented in the figure 1:

- subjective predictions of possible outcomes
- the social and cultural context in which risk is experienced
- the mental images the risk invokes
- the perception of players who are involved in the risk situation and
- the judgements about fairness an equity related to the distribution of potential hazardous events.

The multi-dimensionality of risks and the 'consistent risk belief system' in which our various beliefs about the nature of risks form common characteristics leads to the public risk perception being quite unanimous in almost all countries in which the issue is studied (Renn 2001). This eases the communication that underlies the policies related to diminishing risks.

## The perspective of risk communication – towards deliberative risk governance

Since decision-making is always about exploiting opportunities, policy analysis and planning also need to deal with issues of opportunity and risk. Risks need to be analysed and communicated to decision-makers. Risk communication can be defined as the *process of communicating responsibly and effectively about the risk factors associated with industrial technologies, natural hazards, or human activities* (Leiss 2004). Risk communication is needed in the assessment and management of risks and it needs to take into consideration the nature of the risks and their context (OECD 2002). Risk communication also involves the broader public; the importance of communication also arises from the number of stakeholders involved. Transparency and sufficient communication are thus prerequisites for the successful coordination of different social actors. An important aspect of this capacity is the way communication fosters trust between the different actors, thus facilitating joint action and implementation of policies.

For the purposes of the argument in this chapter, it is interesting to note that an interest towards the context of knowledge is also visible in recent paradigms of risk management, which call for a participatory procedure, involving different stakeholders early in the risk analysis process to characterize risks even before they are given a formal assessment. This does not diminish the role of modelling and quantification, but is aimed at eliciting the values and the perspectives of the community involved so that the multiple dimensions of risk can be taken into account early on in the assessment. (Amendola 2002.)

The broadening view of risk management or risk *governance* (Renn 2007), which has introduced stakeholders into the science-policy equation, implies that the decision-maker does not act alone. In addition, the broader picture with many stakeholders, interest groups and institutions means, in effect, that the moment where the decision is made becomes difficult to pinpoint. The complexity of policy-making leads, thus, to the critique of so-called 'decisionism', which is based on the simplistic idea of a unitary decision-maker making a single decision. (Majone 1989.) Consequently, network identification can be seen as a central concern for successful risk communication, especially in so called wicked problems like climate change, where multiple actors are inevitably addressed (Lehtonen & Peltonen 2006, Schmidt-Thomé & Peltonen 2006). Schmidt-Thomé and Peltonen (2006, 54–55) state that networks of actors, including the resources, capacities and knowledge embodied in the network, may prove invaluable in efforts addressing climate change. Risk communication may be seen as a deliberate attempt to structure and motivate such networks.[6]

Arguing for a participatory and deliberative approach to risk governance, Ortwin Renn has noted the difficulty of combining (scientific) expertise and (public) participation in risk management efforts, and calls for a coherent risk governance framework to facilitate deliberation on risks and risk management (Renn 2001, 2004; 2007.)[7].

Renn (2004) has used the idea of social subsystems to make sense of complex risk governance efforts. Each social subsystem has its specific codes for rendering meaning to societal events. He describes the interacting systems of society with their own operating logics. The economic system, the expert system, the social system and the political system operate with their own respective goals: efficiency, competence in knowledge, fairness and legitimacy. Each of the subsystems, thus, has its own 'demands' which need to be answered in policymak-

---

6 This said, it is also important to recognise that the events themselves have effects on the networks. A single extreme event, such as a flood or storm surge, can give rise to emerging networks which may prove beneficial over time, and thus remain operative. This means that singular events may attain the character of 'focusing events' (Birkland 1997), which serve to mobilize key actors and thus influence the policy agenga, or, perhaps even altering the policy landscape in a more substantial way. The networks seek to remedy the disrupted social practices and, perhaps, invent novel future adaptations to similar events, thus responding to the 'politics of nature', i.e. the disruptive effects and challenges the events pose on prevailing social practices and norms (Haila & Lähde 2003).

7 Renn's (2007) framework seems very useful as it presents a practical model of deliberation, moving beyond the theoretical debate on the virtues and vices of deliberative, participatory planning and consensus-building.

ing. Furthermore, the evaluation criteria of all the subsystems must be addressed, not merely relying on the expert system (i.e. testing truth claims).

The division of labour and the requirements of the subsystems is depicted in Figure 2 (below), which also includes elements of interaction between the four subsystems. In this context, science is filtered through an expert system that contributes to the function of knowledge production through testing truth claims. As to risk management decisions – such as planning in the face of climate change – all four systemic requirements need to be covered; sound policy-making needs to be based not only on solid research, but also economically efficient, effective, politically legitimate (hence the importance of participation and democratic procedures), and socially acceptable.

Fig. 2: Societal subsystems and their logics in decision-making

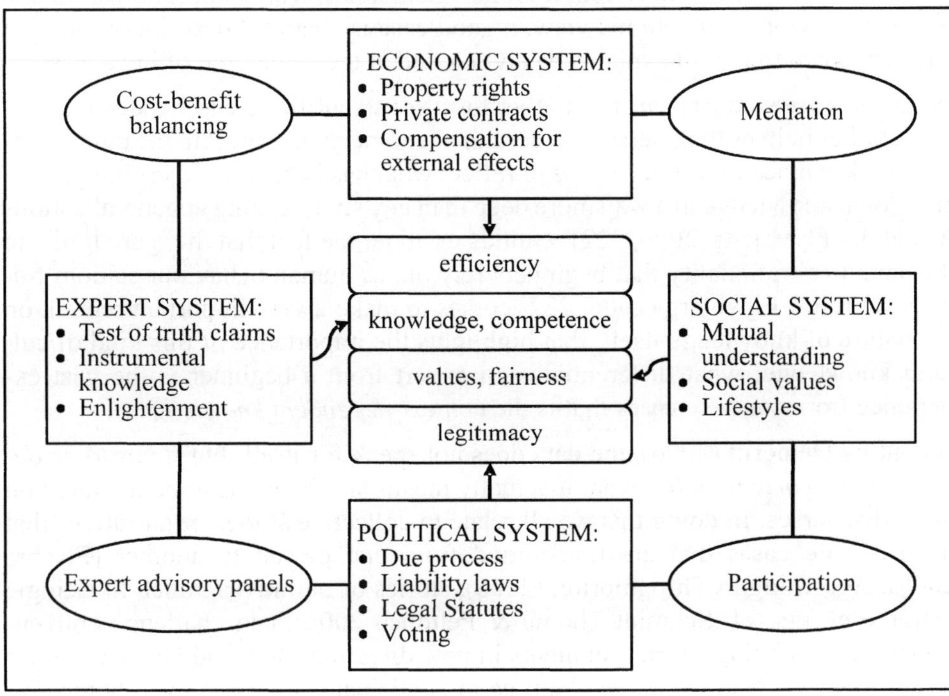


Source: Renn 2004; 295

## The characteristics of case studies

Having presented the theoretical grounds for a more contextual approach to climate change research – and specifically to research on adaptation, we shall now present the case for the case study.

Yin (1984) defines the case study as, „an empirical enquiry", which has the following qualities: 1) It investigates a contemporary phenomenon within its real-life context; where 2) the boundaries between phenomenon and context are not clearly evident; and 3) multiple sources of evidence are used. Other definitions of a case study stress the social and active component in the studies. According to Peuhkuri (2006), referring to definitions given by both Yin (1984) and Stocker (1991), a case study is

> „an empirical enquiry which attempts to understand or explain holistically dynamics of a certain historical period of a particular social unit within its real life context".

Haila (2006) argues that case studies can be understood as on the same lines, stating that a good analogy (as case studies can be understood) is a model for *doing* something and a description on how something works out (and *where* and *when* etc. it works out). In his view, a good case is a careful description of what has happened, that both explains behaviour and at the same time offers a lesson.

What sets case studies apart from other forms of scientific research indeed is, that case studies help us to understand a certain phenomena in respect to the unique setting it takes place in. It thus lets us to reflect what has happen in terms of our own emotional knowledge in a way more deep than any study aiming at generalisations would do. Flyvbjerg (2006, 222) reminds us about the fact that there are limits to the analytical rationality that beginners rely on, as human behaviour seldom follows set rules. Flyvbjerg (2006, 222) comes to make a very important remark on the nature of knowledge itself, that highlights the importance of this kind of cultural knowledge: what differentiates an expert from a beginner is the vast experience from concrete cases that is the *context-dependent knowledge*.

As put by Demeritt (2006), the data does not speak for itself, but *it has to be decided how to orient on it* – what it actually means and how it is to be mounted on other discourses. In doing this we all relate to collective stories or narratives that in effect are cases that are transmitted from one person to another (Fischer 2006). According to Throgmorton (2006), stories or narratives, much like single extreme events (cf. Schmidt-Thomé & Peltonen 2006), can challenge conventional understandings, turn arguments in new directions and find better ways for living together. Narratives also help us to „anticipate situations even before we encounter them, allowing us to envision alternative futures" (Mattingly 1991; 237, cit. Flyvbjerg 2006). The task of the researcher could therefore be seen as providing with scientifically weighted narratives, on the basis of which the warrant for action could be reconsidered through public discussion.

In the light of the ideas introduced above, it seems that disseminating scientific results can not be refrained solely on network building between the scientists and various actors. Decisions on policies leading to action are a result of a complex decision making and public approval process, where the values of the public are repeatedly contested. This applies to climate change adaptation, too, where the risks seen important to adapt to regionally are never the result of scientific analysis alone. Although the importance of one's own experience in building the emotional knowledge our perceptions are based on can not be highlighted too much, there seems to be room for developing local narratives that justifiably contest our views.

Finally, it should be kept in mind that local actors, too, are individuals whose perceptions of risk to large part are based on emotional knowledge (Fischer 2006). Throgmorton (2006) is on the same lines, as he stresses that the personal characteristics and emotions in real world practice become more important than purely disciplinary knowledge. The actors are themselves immersed in cases they act in, so that humanities are in play. Their visions are affected by the politics of expertise, where scientific results are interpreted and modified through their own (hidden) perceptions. This should not be forgotten when we study the actions of different actors (Throgmorton 2006).

## Example: The ASTRA Winter Storm Case Study

In the ASTRA –project case study methodology was used to study the impacts of the winter storm Gudrun, that affected the countries bordering the Baltic Sea (BSR, Baltic Sea Region) on the 7–9$^{th}$ of January 2005.[8]

The aims of the ASTRA winter storm study (WSS) were the following: 1) to make climate change concrete, by asking what could climate change look like in the Baltic Sea Region; 2) to look at climate change adaptation from present-day climate variability perspective, thus bracketing the issue of the future climate; 3) to demonstrate both shared and specific challenges of climate change adaptation in the Baltic Sea Region by studying how a single storm leads to different outcomes in different local and regional settings; 4) to use the case study for drawing lessons on climate change adaptation.

The Winter storm study was based on a questionnaire sent to project partners in different BSR countries to map the impacts of the storm and measures taken in the corresponding countries, on both regional and national levels. Through this stage we were able to identify the most critical impacts of the storm and the exposure units most vulnerable towards it. Later, the study was deepened by focus-

8 The study, titled „Impacts of winter storm Gudrun of 7th–9th January 2005 and measures taken in Baltic Sea Region", or Winter Storm Study (WSS) for short, can be downloaded at www.astra-project.org. In addition to the study summing up BSR level results, a study on the effects of the winter storm in Estonia was conducted by Kont et al. The Estonian study can be found at www.tlu.ee/geo2/astra/GudrunEst%2031.01.pdf

ing on public reports and studies published on these most critical themes mentioned, through internet and library research.

Relatively little effort was spent discussing the specific relation between climate change and winter storms. This link has been widely established. For instance, Münich Re, one of the world's leading reinsurance companies has consistently recognised a link between climate change and storm risk in Europe. As the bottom line of a report on the 1999 winter stroms they have stated the following;

> "It would certainly be exaggerating to develop horror scenarios for Europe as a result of climate change [...] However, there is no doubt that in a warmer climate it will be necessary to expect much more frequent and more intensive windstorm and severe storm events" (Münich Re, 200169)[9]

Thus, for the purposes of the ASTRA winter storm study it was sufficient to note that there is a link between storminess and climate change, as warmer winters come with an increased risk of winter storms. This said, it was more important to focus on local and regional impacts of the storm and lessons learned through response measures.

In total, the storm caused insurance damages estimated to be worth some 1,5 billion euros and led to 17 casualties. Hundreds of thousands of people were affected by the storm. The impacts of the storm were manifold; on 7th of January 2005, floodwaters poured into the streets of Pärnu. A few hours later, the city of Helsinki was struggling with the highest sea-level ever, city rescue department mounting cubes of recycled paper on the sides of the harbour pool to form an improvised flood-wall. The Loviisa nuclear power plant on the Finnish coast nearly came to a halt, as the rising sea-level reached levels critical to the plant's cooling water system. In Denmark, the storm stalled thousands of wind power units while in Sweden, it felled timber totalling to an amount comparable to a year's cuttings. Overall, Sweden suffered the largest economic losses in the Baltic Sea region.

The study showed many interesting aspects about the current state of adaptation and of the networks of actors contributing to dealing with local vulnerabilities. Although meteorologically a single event, the impact of the January 7–9 2005 storm varied greatly in intensity and nature among the BSR countries. As did the impacts, so did the responses. Adaptation to the impacts of extreme weather events, be it anticipatory or reactive, was in action seen to depend greatly on the characteristics of each area in question. In many cases, like in Helsinki, where based on earlier experience on similar flood situations, ad hoc flood defences were set in place to protect the historical centre, the adaptive actions were the

9 It may be noted here that the societal interpretation and relevance of environmental problems such as climate change, is constructed through crucially depends on actors such as the insurance industry who have important interests to protect and are thus keen to gather information and act upon it.

result of individual actors reading the situation right and turning their expert knowledge into action.

Taking this notion by heart, we come to praise the importance of in-depth studies on exemplary cases like the Winter Storm Study; We are not merely stating the outcome of a single storm event, we are *studying* the impacts of the extreme event in a certain *context* that is the socio-economic and natural setting of the area, in order to *learn* from the event and ultimately *pass the information forward* to the relevant actors. For example, according the WSS, the long-lasting power outtakes in Denmark, Sweden and Latvia all had very different reasons. In Denmark, where a relatively large share of energy is produced by wind turbines, the shut-down of some 4000 turbines because of high winds was the main cause. Although the plants operate automatically, interestingly, the situation was made worse by the human component – some older turbines have to be manually started after a halt, and because of the storm hitting on weekend not all the persons responsible were available. This prolonged the electricity cut.

In Sweden the blackouts were caused by trees fallen on electricity lines. The situation was made worse by the huge amount of fallen timber on the roads that slowed down the process of reaching and fixing the damaged cables. The problem itself was caused partly by local factors, as the distribution of tree species in Sweden's forests is inclined towards spruce monocultures that are vulnerable for storm damages. Meteorological conditions other than the wind speed also played a role; as the winter was exceptionally mild, ground frost that normally supports the trees was missing. As primary concern was to clear the main ways, sparsely populated areas had to wait for electricity for days at worse.

As key results it was seen, that immediate reactions to the storm were mostly reactive, as were most of the improvements to the governance processes reported by our project partners. Only in Sweden and Latvia were excessive improvements made, in the form of improving coordination and dividing responsibilities between different actors and building communication networks more robust. Just as valuable as these 'visible' improvements though, it can be argued, are the new, even though temporary networks the event created.

As Schmidt-Thomé and Peltonen (2006, 54) state, many actors and structures had to function together to form a network, as the measures taken simply had to be taken. In many cases institutional codes of conduct exist, but just as often actors have developed local solutions for reactive adaptation themselves. This was perhaps best seen in the construction of flood defences in Helsinki. Local knowledge is essential, in order for both the local actors adapting to climate change and for the scientists studying the possibilities for adaptation to be experts in their field. To maintain this expertise, in-depth knowledge of cases is needed.

The Winter storm study has been used during the ASTRA project as a starting point for awareness-raising on climate change adaptation issues. Presentations and dissemination of the case study have drawn attention to the effects and extent of the storm, also pointing to the number of affected people and key stakeholders whose future actions may mitigate the impacts of future storms. The ASTRA winter storm case provides a starting point for future efforts in the direction of case-based learning. Case-based learning has been used as a method in engineering, medicine and social sciences (e.g. Ruggiero 2002; e.g. Prince & Felder 2006). Where traditional approaches of teaching have been deductive, proceeding from the general (universal law) to the particular (instances, cases), the case-based approach is more inductive, starting from the particular and building towards a broader picture. Prince and Felder (2006) point to consistent research findings suggesting that inductive methods are in general more effective than traditional deductive methods for achieving a broad range of learning outcomes.

## Instrumental, critical and persuasive aspects of case studies

The January 2005 storm was exceptionally strong, but not surprising as such. Extreme weather events in some form already affect all of the communities bordering the Baltic Sea, thus manifesting their current vulnerability towards climate-borne effects. Due to its exceptional character, the 2005 storm provided insights into the processes relevant for climate change adaptation in the BSR. The winter storm revealed things about how the policies and networks of actors dealing with the adverse impacts of climate change should be refined in terms of adaptive capacity towards extreme weather events. This aspect made the WSS *an instrumental case study*, i.e. a case that serves in the development of more general themes and theories. In an instrumental case study however, less weight has to be put on this, as the aim of the study is to find generalisations between the case (e.g. winter storm of January 2005) and the issue under investigation (e.g. climate change adaptation). However, to understand the event thoroughly enough in its context is a precondition for drawing conclusions on how similar events might affect the issue under study. (Stake 2003, 3).

In other words, the case had the potential to become a reference point of the current vulnerabilities to climate-borne effects on the BSR. The ASTRA-project aimed at a holistic view on the impacts of a single storm event on the countries of the BSR', showing the multiplicity of local impacts and responses. The study thus gave an impression on what issues might be relevant in terms of the impacts of climate change for further study in the project. Furthermore, it was to serve as a basis for conversation among different stakeholders, on how to pursue the somewhat abstract and new issue of climate change adaptation.

By another definition, interestingly, the WSS can indeed be seen as both *an extreme case and a critical case*; in the current context, the storm event was as an extreme event. Thus it served as a good case by potentially offering greatest possible amount of information on the phenomenon at hand, by engaging maximum amount of actors and activating more basic mechanisms than a normal event would have done. The Winter Storm Study aims more to 'clarify the deeper causes behind a given problem and its consequences than to describe the symptoms of the problem and how frequently they occur', thus making it *an extreme case* (Flyvbjerg 2006, 229).

Since the purpose of the WSS was to give an insight of the impacts and means of adaptation to climate change, the study can also be seen as *a critical case* (Flyvbjerg 2006). While being extreme in today's climate, a similar storm according to studies might well be considered 'normal' in the future, referring to the intensity and theoretical recurrence of the event. Therefore, the capacity to cope with such an extreme event today can be seen crucial for overcoming tomorrow's expected challenges. If the means of reducing vulnerability to current climate-induced effects are not adequate, it can be seen that they will not be adequate in the future either.

The Winter Strom Study, by the nature of case studies, is finally a narrative presented by the researcher. Therefore, it is only one version or interpretation of the truth, and actors can and, indeed, should read it through the scope of their own experience and values. The impacts on different exposure units presented should be seen as examples on possible effects only. The actors interested on the possible impacts of extreme weather events on their own areas have to interpret them in the context of their knowledge of the local socio-economical, geographic and even political features (Stake 2005, 43).

The instrumental and critical case study provides the possibility to 'ring the bell' about an important issue, such as climate change adaptation. This reveals yet another, persuasive aspect of the case study. In instances where a new policy issue is at hand but is not receiving enough attention, and is being pushed on the agenda by knowledge brokers, knowledge production is not only about neutral information, nor is it strictly about advice; rather, this kind of information is akin to *persuasion*. As Majone (1989) argues, a persuasive approach to knowledge production is needed when the policy analyst's most difficult task is not to answer what should be done, but how to make policy-makers listen. This is due because decision-makers usually operate with traditional categories, restricted to a stable set of objectives and means to achieve them:

„Persuasion is needed to induce them to consider different formulations or approaches to the issue under discussion since the psychological effect of factual arguments may not be strong enough to overcome the inertia of long-established patterns of thought. Experience also shows that facts and statistics are seldom sufficient to bring about changes in behaviour – even after the need for change has become clear" (Majone 1989, 34).

This is why concrete case studies are important; they are factual but go beyond facts and figures in their narrative form. First, case studies can be instrumental in evoking a bigger picture through a easily understandable small-scale event. Second, they may have extreme or critical characteristics, pointing to real-life 'worst-case-scenarios'. Third, partly as a consequence of the above features, case studies have persuasive power, potentially breaking the habitual thought-patterns of decision-makers and the broader public.

## Conclusions

We have grounded our argument emphasizing the utility of case study methods in climate change adaptation research in three theoretical 'movements': 1) the epistemic movement away from the quest of certainty, favoring a pragmatist approach; 2) the move in policy analysis from unproblematic 'policy advice' towards interpretive and contextual narratives, grounding policy formation and evaluation in a critical consideration of social values and social relevance, and; 3) the move in risk communication from monologue to dialogue and towards risk *governance*, a framework of participation and deliberation where

The three perspectives provide theoretical grounds for the use of case studies in climate change adaptation research, which may help actors-in-context to devise adaptive measures based on analogical accounts and grounded narratives by actors in similar or comparative localities. We have stressed the importance of in-depth studies on exemplary cases such as the Winter Storm Study. The winter storm study, even if it is not the best example of 'thick description', demonstrates the potential of a *critical* and *persuasive* case study in speaking to the development of local practice and policy development of climate change adaptation. The case study was not a bookkeeping effort on the outcome of a single storm event; rather, it was a study that articulated the link between climate change and local extreme events by studying the impacts of the event in a specific *context* (the Baltic Sea region) that is the socio-economic and natural setting of the area, in order to *learn* from the event and ultimately *pass the information forward* to the relevant actors.

Under ASTRA it was seen that a single case study could tell a lot about the current state of policy making in the Baltic Sea area, and on actual responses for reducing local and national vulnerability towards weather related extreme events. The study, and other similar studies could well be used as starting points for local and regional workshops where the particulars of the winter storm in that location and local institutions with local actors (and their narratives!) are taken on board to engage in more in-depth sense-making efforts of the local events in January 2005.

Although the lack of relevant policies was in some cases evident, this lack was to some extent balanced by the action and learning process that the event initiated. Potentially, a single event can have effects on lengthy policy processes, the results of which can not be seen immediately after the event. Examples of such „instigating events" can be found at different scales (Consider, for instance, the Swedish decision to phase out nuclear power after the Chernobyl accident in 1986). Likewise, at the local level, individual environmental issues and conflicts may have long-lasting effects on the local political agenda (Laine et al. 1998). The locally grounded narratives on the impacts do help to make abstract issues visible however, and in the long run surely help to appreciate the importance of adaptive actions. As events like Gudrun repeat themselves, as they repeatedly did in January 2007, they slowly take their part in reshaping our emotional knowledge and are a basis of a new type of grounded narratives that slowly re-form our values.

## References

*Adger, W.N., Brown, K., Fairbrass, J., Jordan, A., Paavola, J., Rosendo, S. & Seyfang, G.* (2004), „Governance for sustainability: Towards a 'thick' understanding of environmental decision making", CSERGE Working Paper EDM 02–04.

*Amendola, A.* (2002), „Recent paradigms for risk informed decision making". Safety Science 40 (1–4), pp. 17–30.

*Birkland, T.A.* (1997), After disaster: Agenda setting, public policy, and focusing events, Georgetown University Press, Washington.

*Demeritt, D.* (2001), „The construction of global warming and the politics of science." Annals of the Association of American Geographers 91 (2), pp. 307–37.

– (2006), Traditions of explanation in environmental geography. Lecture given at the YHYS conference 2006, Tampere, Finland 23.11.2006

*Demeritt, D. & Langdon, D.* (2004), „The UK Climate Change Programme and communication with local authorities", Global Environmental Change 14 (2004), pp. 325–336.

*Dewey, J.* (1929), The Quest for Certainty: A study of the relation of knowledge and action, Minton, Balch and Co, New York.

*Dessai, S., Lu, X. & Risbey, J.S.* (2005), „On the role of climate scenarios for adaptation planning", Global Environmental Change Part A 15 (2), pp. 87–97.

*Easton, D.* (1953), Political system, Knopf, New York.

*Eboreime, T. & Wennerström, P.* (2005), Ilmastonmuutos. Kansalaisten tiedon taso ja arvoryhmätarkastelu, Taloustutkimus Oy, Helsinki. 32 p. Available at: www.ilmastonmuutos.info/fi/cfmldocs/document.cfm?doc=show&doc_id=218

*Fischer, F.* (1995), Evaluating Public Policy, Nelson-Hall Publishers, Chicago.

– (1998), Beyond empiricism: Policy inquiry in pospositivis perspective. Policy Studies Journal 26:1, 129–46.

– (2006), Citizens and Experts in Environmental Struggles: Knowledge in Situational Context. Lecture given at the YHYS conference 2006, Tampere, Finland 24.11.2006.

*Fischer, F. & Forester, J.* (eds.) (1993), The Argumentative turn in policy analysis and planning. Duke University Press, Durham and London.

*Flyvbjerg, B.* (2001), Making Social Science Matter. Why social inquiry fails and how it can succeed again, Cambridge University Press, Cambridge.

– (2006), „Five Misunderstandings About Case-Study Research“, Qualitative Inquiry, vol. 12, no. 2, April 2006, pp. 219–245.

*Haila, Y. & Lähde, V.* (2003), Luonnon politiikka, Vastapaino, Tampere.

*Hajer, M.* (2001), „The Need to Zoom out: Understanding planning processes in a post-corporatist society." In A. Madanipour, A. Hull, P. Healey (eds.), The governance of place: Space and planning processes, Ashgate, Aldershot. pp. 178–202.

– (2002), „Setting the stage: A dramaturgy of policy deliberation", Administration and Society 36(6), pp. 624–47.

*Kunelius, R.* (2003), Viestinnän vallassa. Johdatusta joukkoviestinnän kysymyksiin, WSOY, Helsinki.

*Laine, M., Peltonen, L. & Haila, Y.* (1998), „Local „Instigating Disputes“ and the Polticization of the Environment: The case of Tampere, Finland." Advances in Human Ecology 7, pp. 161–206.

*Lehtonen, S. & Peltonen, L.* (2006), „Risk communication and sea level rise : bridging the gap between climate science and planning practice“. In Schmidt-Thomé, P. (ed.), Sea level change affecting the spatial development of the Baltic Sea Region, Geological Survey of Finland Special Paper 41, Geological Survey of Finland, Espoo, pp. 61–69.

*Leiss, W.* (2004), Effective risk communication practice. Toxicology Letters 149 (1–3), pp. 399–404.

*Majone, G.* (1989), Evidence, argument and persuasion in the policy process, Yale University Press, New Haven.

Münich Re (2001), Winter storms in europe (II): Analysis of 1999 losses and loss potentials. Münchener Rückversicherungs-Gesellschaft, Munchen.

OECD (2002), OECD Guidance document on risk communication for chemical risk management. OECD environment, health and safety publications. Series on risk management no. 16. Organisation for Economic Co-operation and Development, Paris.

*Nowotny, H.* (2000), „Re-thinking science: From reliable to socially robust knowledge". In M. Weiss (ed.), Jahrbuch 2000 des collegium helveticum, Zürich, pp. 221–244.

*Peltonen, L., Haanpää, S. & Lehtonen, S.* (2006), Extreflood – Tulvariskien hallinta yhdyskuntasuunnittelussa. (Flood risk management in urban planning). Suomen ympäristö, luonnonvarat 22/2006, Finnish Environment Centre, Helsinki. Available at: www.ymparisto.fi/default.asp?contentid=188151&lan=fi

*Prince, M.J. & Felder, R.M.* (2006), „Inductive teaching and learning methods: Definitions, comparisons, and research bases" Journal of Engineering Education 95 (2), pp. 123–138.

*Renn, O.* (2001), „The need for intergration: risk policies require the input from experts, stakeholders and the public at large", Reliability Engineering and System Safety 72 (2001), pp. 131–135.

– (2004), „The challenge of integrating deliberation and expertise". In T. L. McDaniels and M. J. Small (ed.), Risk analysis and society. An interdisciplinary characterization of the field, Cambridge University Press, Cambridge, pp. 289–366.

– (2007), Risk governance: Coping with uncertainty in a complex world, Earthscan, London.

*Rittel, H.W.J. & Webber, M.M.* (1973), Dilemmas in a general theory of planning. Policy Sciences 4 (2), pp. 155–69.

*Ruggiero, J.A.* (2002), „'Ah ha' learning: Using cases and case studies to teach sociological insights and skills. Sociological Practice 4 (2), pp. 113–28.

*Schmidt-Thomé, K. & Peltonen, L.* (2006), „Actors, networks and actor-networks in coping with sea-level rise". In Schmidt-Thomé, P. (ed.), Sea level change affecting the spatial development of the Baltic Sea Region, Geological Survey of Finland Special Paper 41, Geological Survey of Finland, Espoo, pp. 51–59.

*Stake, R.E.* (2003), The Art of Case Study Research, SAGE Publications, Thousand Oaks.

*Suhonen, P.* (1994), Me, mediat ja ympäristö, Hanki ja jää, Helsinki.

*Throgmorton, J.* (2006), The Bridge to Gretna: Three Faces of a Case. Lecture given at the YHYS conference 2006, Tampere, Finland 23.11.2006

*Weimer, D.L. & Vining, A.R.* (1992), Policy analysis. Concepts and practice (Second edition), Prentice Hall, Englewood Cliffs.

*Yin, R.K.* (1984), Case Study Research. Design and Methods, SAGE Publications, Newbury Park.

# Chapter 7
# Educational Actions in Italy to Face the Problem of Climate Change

**Eugenia Aloj, Mariagrazia De Castro, Anna Zollo**

## Abstract

This paper deals with the climate change in Italy and the environmental information, education and communication of the topic in Italy.

After presenting a general picture of the activities at the institutional and non-institutional level, we switch focus to the role of the scientific community. We then continue to focus on our role as research staff involved for many years in environmental education and education for sustainable development. Hence, we show our activities, highlighting the integrated goals and the connection between these and the tools used. This connection plays a strategic role in talking about environmental education for climate change as life-long learning.

## Climate change as global change

Climate changes are considered one of the most imminent threats to the health and wellness of mankind, the economy and world environment.

Also UNESCO has said that global warming is a "modern" complicated problem, involving the entire world, tangled up with difficult issues such as poverty, economic development and population growth.

The scientific international community is in agreement that the world's climate has been influenced by gas emissions from economical activities such as industry and transports that produce carbon dioxide, methane and so on. At the international level, climate changes have been addressed since 1992 when the Convention for Climate Changes of United Nations (UNFCCC) was defined to have an organic picture of those initiatives oriented around reducing gas emissions. Since 1997, the international community has addressed the problem in a more concrete way, assigning Kyoto's protocol to committing developed countries and transition economy countries to reduce gas emissions by 5% within the period 2008–2012. Specific goals of reduction are different for each country. The European Union is committed to a reduction of 8% of emissions within the aforementioned period. Subsequent agreements differ and have redistributed this commitment so, to ratify them; Italy has to reduce its own emissions by 6.5%. Our country is one of the nations included in Annex B of the Kyoto Protocol. By law in 2002, Italy ratified the protocol and in the same year approved a National

Programme for the reduction of gas emissions. In 2004, a new law regarding sharing the disposal of emissions began with collaboration with Mexico involving the transfer to our country of 5 million tons of carbon dioxide. Only in 2005 was the protocol really still in force.

Fig. 1: Clean air for the planet's survival.

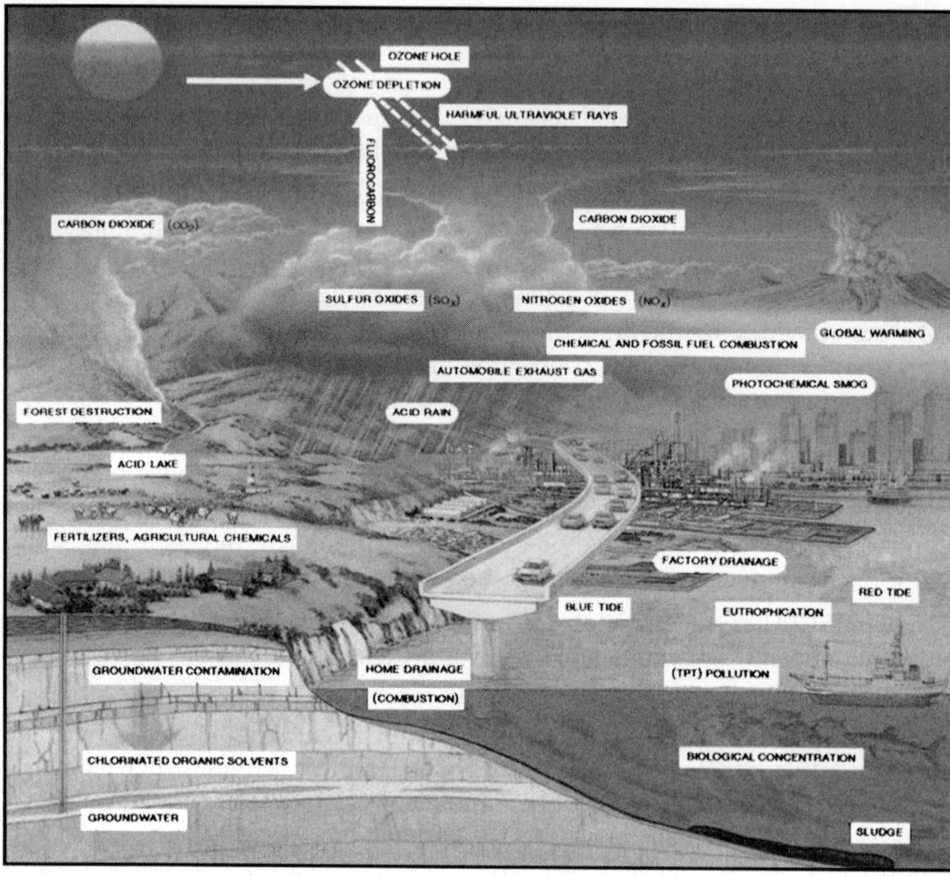


Source: Shueisha (Imidas 1990)

## Climate change in Italy - the current situation

Data from the Ministry of Environment shows the current situation for emissions in Italy as follows:

Tab. 1: Italian scenario by 2010 for gas emissions

| **Mt $CO_2$ eq** | **Year 2000** | **Year 2010** |
|---|---|---|
| **Fields of emission** | | |
| **From energy sources:** | 452.3 | 484.1 |
| Energy industry | 160.8 | 170.4 |
| Building industry | 77.9 | 80.2 |
| Transports | 124.7 | 142.2 |
| Services and Public Administration | 72.1 | 74.1 |
| Agriculture | 9.0 | 9.6 |
| Other (distribution services, military services, and so on) | 7.8 | 7.6 |
| **From other sources:** | | |
| Industrial processes (chemistry) | 33. 9 | 30.9 |
| Agriculture | 42.6 | 41.0 |
| Waste | 14.2 | 7.5 |
| Other (solvents, bunkers) | 3.8 | 16.7 |
| **TOTAL** | **546.8** | **579.7** |

This trend scenario shows a negative situation in which further reductions are very difficult. Italy is in a difficult position to reach goals and commitments of Kyoto. $CO_2$ emissions have increased from 1990 to 2002 by about 10%.

Mainly, increased emissions are from the electricity energy field and from transport causing energy policy decisions that are very far from real attention to Kyoto and its commitments.

Therefore, it seems that there is a lack of environmental education.

Our effort in order to get an answer to this lack of environmental education has been a general survey about environmental education, dissemination activities about climate change in Italy strongly oriented to find project elements in its activities and then some proposals for improvement and interventions.

Since beginning our survey, we have identified that environmental education activities about climate change are mainly done by:

1. Institutional subjects (universities, Ministry of Environment, public agencies on the topics, local public agencies);
2. Non-institutional subjects (non-profit organisations, private companies).

The outcome of our survey is a general picture of 2006 regarding the main educational and dissemination interventions on climate change:

3. The survey: institutional and non-institutional level – a general picture.

## Institutional Level (national)

1. UNESCO: in the context of Decade of Education for Sustainable Development the National Week for Education to Sustainable Development is wholly dedicated to energy. The theme has been chosen by the National Italian Commission for UNESCO during the annual meeting of participating agencies: regions, NGO, research centres, associations, and school networks, in consideration of the large worldwide interest into the research of new solutions for the reduction of energy consumption.
2. Ministery of Environment: Minister of Environment considers "climate change" as a key point in the education of and for the level of awareness of the public.
3. ENEA: (acronym for National Agency for Energy and Environment) has a department dedicated to climate change that undertakes various research activities and focuses on educational goals which it supports with studies and surveys, and manages environmental and energy data of the activities of involved agencies and companies.
4. Universities: We focus on the activity of the Italian Society of Ecology (Site) that began in 2001 when a public letter was written to the premier highlighting Kyoto in the perspective of a new political deal for the reduction of emissions. In September 2006, a national congress was organised regarding global changes, ecological diversity and sustainability as three integrated topics. Regarding Prin, (acronym of national interest research projects) by universities, we have found only one Prin on the topic studying the impacts of climate change on the degradation of atmosphere and environmental implications on high mountains (by University of Studies of Milano).
5. National and regional agencies: APAT, the National Agency for protection of environment deals, has for many years, focused the topics of climate change on a web page, "Protection of the atmosphere at global level" and in a project about environmental quality in metropolitan areas. There is a section dedicated to "air emissions" and air quality which focuses on the monitoring of Pm levels in the metropolitan Italian areas. APAT also has disseminated pamphlets on the topics as thematic schedules as basic information for people. Regional agencies of APAT are involved in the topics. For example, in Piemonte, there was a scientific course and laboratory about climate change organised last September in the town of Turin.
6. Municipalities and their work on the topic are expressed under Agenda 21. We have included Agenda 21 for the evaluation of activities made by municipalities where climate change processes have a strategic role. Therefore, in order to have a general survey, some data from 2004 is shown where climate change is contemplated in an extended way in the field of energy.

Fig. 2: Items of Agenda 21 in 2004

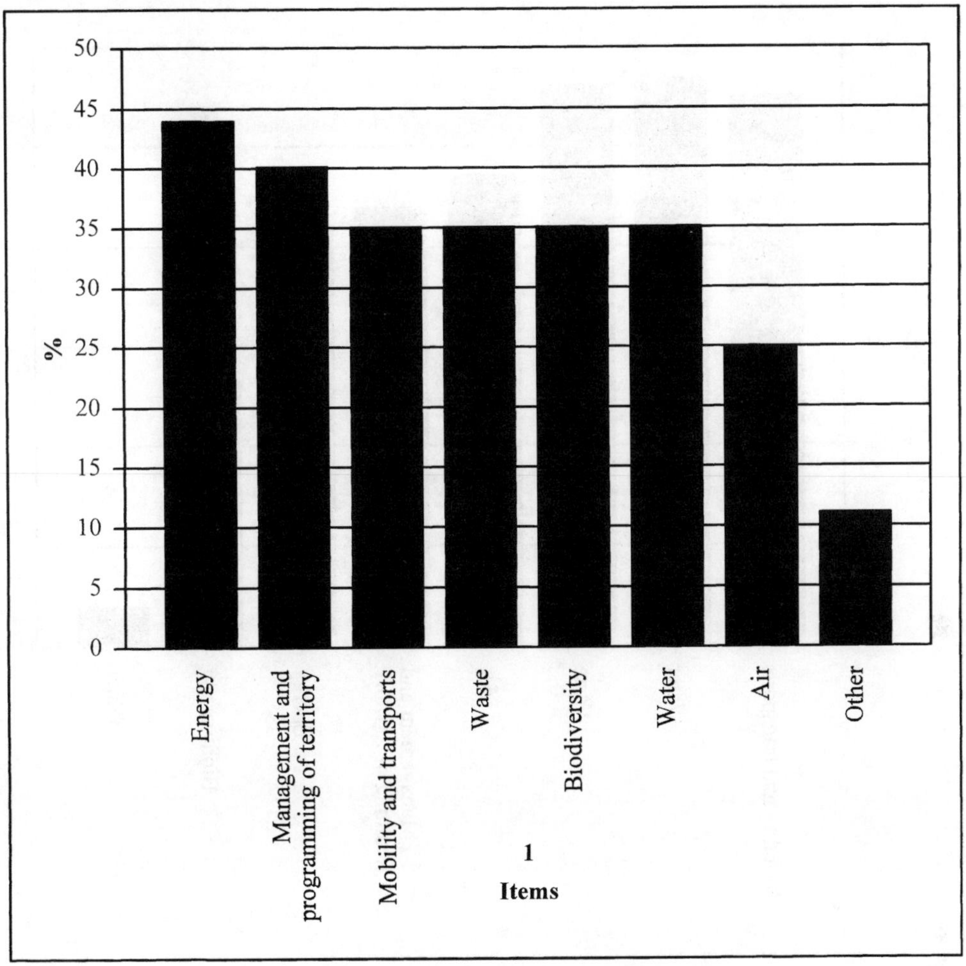


In Table 2, areas where municipalities have implemented Agenda 21 are shown and we find that the most disseminated item where climate change is comprehended is Energy. (Note that the total amount in percentage is more than 100% because there was the possibility of more than one answer).

Fig. 3 Problematic areas in Agenda 21

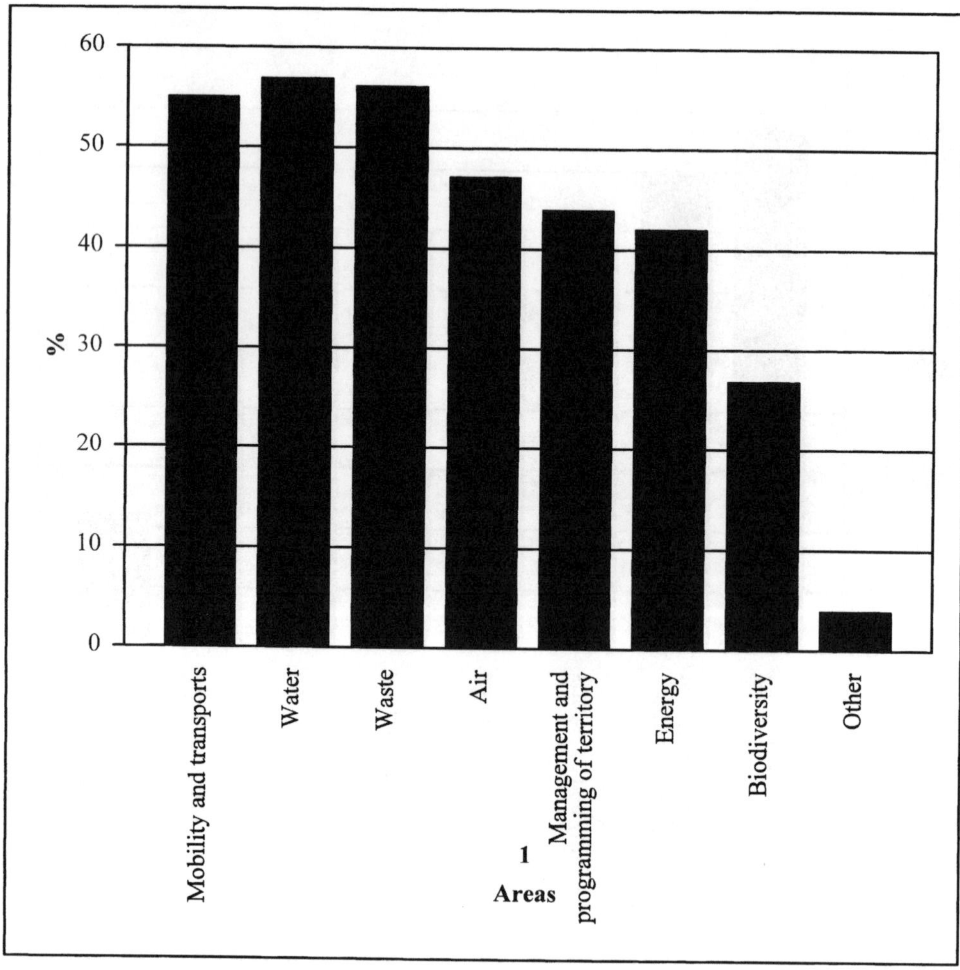


In Table 3 we show that the most problematic areas to deal with are Mobility and Transports which, as we know, are the most important sources of atmospheric pollution and consequently of climate change.

Other important activities undertaken by municipalities are expressed as energy saving management with the potential of environmental education. One of these is the institution of PEC (acronym for municipal energy planning). This is a tool linked with other local strategies of planning and of sustainable development and is used by public administration to empower the development of renewable sources, the promotion of an intelligent use of energy and to adjust building to reduce energy consumption.

Normally a PEC is a document composed of various parts in which the energy sources and the consumption of a town are described and focuses on an environmental balance account between consumption and sources concluding with strategies for reduction in consumption and improvement of renewable resources.

A law made in 1991 establishes that the general planning of municipalities must include a specific energy plan if the town has more than 50,000 inhabitants. The same law has introduced the position of an Energy Manager for those industrial companies that have an annual consumption of more than 10,000 tep and for public administration of more than 1000 tep. The role of this manager is to intervene to reduce energy consumption.

7. Coldiretti: Another important role, at the institutional level is made by Coldiretti, the National Agency for Agriculture. Concurrent with the recommendations by Kyoto, the reduction of petrol and gasoline use by the substitution of biofuel promotes the cultivation of plants, for example, sunflower oil, providing new sources of energy. They estimated that the utilization of 20–30% of total useful agricultural surface could contribute to the reduction of 5–6% of total petrol consumption, also providing new jobs.
8. ENEL, the national agency for the supply of electric energy, at the educational level, Enel is very involved with various activities:
   - a game combined with a national award to involve schools to comprehend the energy condition of their territory;
   - a virtual tour in a centre for production of energy for secondary schools to comprehend the complexity of production;
   - nature routes for all the twenty Italian regions presenting the energy landscape for tourist activities;
   - a video about the history of electricity.

## Institutional Level (Regional Level)

9. Activities undertaken by Campania Region. In Campania Region ARPAC, the regional agency for environmental protection, has instituted CRIA (acronym for Regional Centre for Atmospheric Pollution) to monitoring and check air pollution and emissions in the air by factories. The centre has activated a network of centre lines disseminated to the regional territory.

   In the context of institutions, the activities carried out by regions, departments and other local agencies are also very beneficial, in particular the Department of Productive Activities in Campania Region. This department has increased funds to companies financially supporting interventions that improve energy effectiveness and save energy. Some financed projects deal with wind energy systems, photovoltaic systems, biomass from wood and liquid waste from zoos, vegetable oils, hydro electric energy and co –generation (integrated generation of energy and warmth from methane). Another financial fund was made to support the production of renewable energy by welcome services, e.g. hotels and restaurants to enable the eco-management of welcome structures. Focusing on environmental education the same department launched the project, "A scuola con energia" (At school with energy). This is organized into two parts. The first one is for elementary schools (6–11 years) using laboratories where the pupils learn by gaming, the second one is an award for middle schools (11–14 years) where the classes are coordinated by a teacher and are involved in the elaboration of a communication campaign (posters, brochures, slogans) with graphic and multimedia tools (spots, short movies) about lowering energy consumption. The project also deals with seminars and other learning activities. An outcome of the project is the website www.ascuolaconenergia.it in which is possible to download materials and interact with experts with the perspective of training new generations.

Tab 2: The evaluation criteria for 2006 in Campania Region.

| **Pollutants** | **Criteria** |
|---|---|
| $SO_2$ | The value of 350 μg/m$^3$ per hour may not exceed more than 24 times in a civil year |
| CO | The average value in 8 hours may not exceed 10 mg/m$^3$ |
| $NO_2$ | The value of 240 μg/m$^3$ may not exceed more than 18 times in a civil year. |
| PM10 | The daily value 50 μg/m$^3$ may not exceed more than 35 times in a civil year. |
| $O_3$ | The value per hour is 180 μg/m$^3$; the alarm is at 240 μg/m$^3$ |

Therefore, the survey has been carried out by monitoring non-profit organisations (we have surveyed organisations of national importance). Monitoring their activities is considered very useful and important in the promotion of the sustainable use of the earth's resources through education and research.

## Non-Institutional Level

1. WWF

WWF Italia has programmes and activities dedicated to climate change. Recently it disseminated a pamphlet entitled "Il mago del clima" (magic of climate) in the context of a whole year of activities dedicated to environmental changes. It is a pamphlet for pupils of participating schools. In the context of one whole year, they organised an *online training course* for teachers focusing on the topics of air emissions to disseminate a new culture of the environment as the main priority for our common future.

2. Legambiente

Legambiente has organised a school award entitled "Scuola amica del clima" (school as a friend of the climate) for all interested schools. The award is subdivided into two sections. The first section is dedicated to actions for energy efficiency and is awarded to schools. The second section is for the best examples of energy saving education and is awarded to schools oriented towards energy efficiency. Another important activity of Legambiente on the topic is the educational project "Cambio di Clima" (Climate Change). It includes online training for teachers and the participation of pupils in a national competition for the best environmental education project. Legambiente provides some resources for the pupils such as a didactic framework on energy and promotional products like journals, cards, stickers, brochures, vademecum, flags and certificates, as well as some resources for teachers such as journals, newsletters, stickers and cards for associated members. Furthermore, in ten Italian towns the activities were supported by Ecoludobus, a bus with a travelling laboratory of environmental education on which pupils play and learn about energy consumption. Finally, Legambiente has organised a training course, "EnergiKamente", on the global energy question. The training is for 8 weeks of study, exercises, preparation of pamphlets, discussions on forum and interaction for teachers of different disciplines, educators, trainers and cultural operators.

In the town of Rispescia, Grosseto Toscana Region, Legambiente has organised the Centre for Renewable Energies, a reference point for the promotion and dissemination of renewable energies. It was born as a point of information for the public and for companies, but also as a point of communication and training. In fact, in the centre there is a permanent exhibit about renewable energy for schools and the public.

3. Greenpeace

Greenpeace Italia's mostly relevant activity deals with a campaign on "Energy and Climate" supporting the following goals for Italy:

- improvement of renewable energies: actually we are at 2.5% of renewable energies of total energy but by 2020 we will have to reach at least 20%;
- improvement of energy effectiveness to reduce the demand for energy by industry;
- growth of investment in research and development of new energy resources.

The synthetic map in Table 3 allows us to examine the role of the scientific community in environmental education and communication for climate change.

Tab. 3: A synthetic map

| **Institutional Level** | |
|---|---|
| UNESCO | National Energy Week |
| Ministry of Environment | Climate change as key role in environmental education |
| ENEA | Department of education and research |
| Universities | Few research projects |
| National and regional agencies | Webpage, pamphlets, section of projects dedicated to air emissions |
| Municipalities | Agenda 21 (climate change, energy, mobility)<br>Energy Municipal Planning<br>Energy Manager |
| Campania Region | "Ascuolaconenergia" and funds for companies |
| Coldiretti | Biofuel |
| Enel | Educational Projects |
| **Non-Institutional Level** | |
| WWF | Pamphlets for pupils, online training courses for teachers |
| Legambiente | Training online courses for teachers, school awards, travelling laboratory, educational projects for schools, stickers, brochures, vademecum<br>Centre for Renewable Energy in Rispescia |
| Greenpeace | Campaign on Energy and Climate |

## The role of the scientific community: focus of our experience

Summarising the points we think are the main causes of climate change:

1. Main use of non renewable energy in all economic and social activities is the production of goods, heating in houses, etc.;
2. Atmospheric pollution from motorized mobility (excessive use of private cars instead of public transport) and from heating of houses;
3. Non sustainable consumptions in home activities;
4. Non sustainable production by industries (unfortunately few industries have energy saving management as a way of managing their relationship with the environment).

The joint research staff at Sannio University (Chair of Ecology in Economics Faculty) and Suor Orsola Benincasa University (Chair of Environmental Education and communication in a Training Sciences Faculty) have been strongly involved for many years in environmental education for sustainable development. We think that the main guideline is a programme of education and communication oriented to the promotion of different integrated goals.

Therefore, what specifically is the role of the scientific community? The role is the strong emphasis on environmental education on the prevention of climate changes. Environmental education could assure a sustainable modification in behaviour as uncorrected models of production and consumption have the effect of altering atmospheric dynamics.

Fig. 4: Framework for our activity of environmental education and communication for climate change.

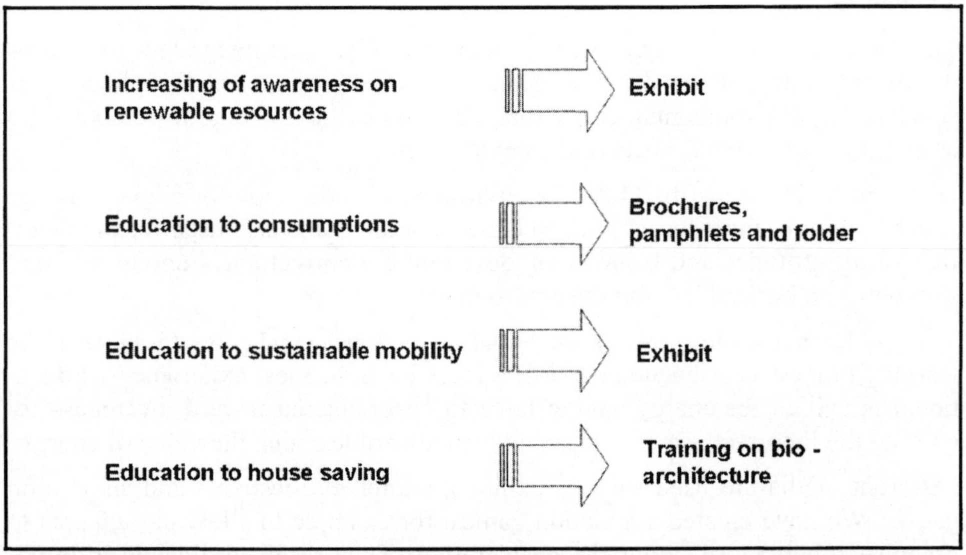


Through this framework, environmental education is able to perceive two goals: promote of awareness of environmental problems and develop quality and dynamic capabilities such as initiative, independence, and availability of commitment and responsibility (Aloj 2000, 41). The first goal is often considered as hostile to economic and technological development, the second one is often considered as a positive attitude for modernity.

For these reasons there is the need for projects based on the awareness that environmental problems and productive qualities are strongly connected but, at the same time, the development of human creativity and human potential is the ideal humus for lifelong results.

Therefore, environmental education projects and activities could be orientated towards increasing the connection between social, economical and cultural topics creating a teaching effort that promotes a method based on the assumption of responsibilities.

This is the key idea on which we are working; developing initiatives of environmental education that is sustainable, focusing on young generations and their emotional participation in improving the environment.

Sustainable behaviour means the acquisition of awareness for pupils and their families about energy saving that in an extended way leads to comprehension of the consumption of electricity, heating in houses, means of transports, etc., stressing acknowledgement of renewable sources like wind, waste, hydrogen, and nuclear.

The different goals and tools we have shown in Fig. 4 are integrated. In fact the use of different tools for different goals is justified in the cross disciplinary perspective of environmental education, deepening the system of relationships amongst the different aspects of the environment.

Different tools are justified because environmental education for climate change is used not only for information and dissemination but is also used for the modification of attitudes and behaviours, developing a correct and lifelong relationship between humankind and the environment.

The problems regarding energy are global. In the European Union it is better if the learning process is orientated towards a local level, to local experiences. Educational activities for energy saving have to be orientated toward awareness regarding the incorrect use of energy but also toward teaching the value of energy.

Different media are used such as exhibits, pamphlets, festivals and simulation games. We have created simulation games, for example to allow participants to organise coherent behaviour in consideration of the received didactic information. Furthermore, another educational tool that we have used is the exhibit of the dissemination of environmental messages to all members of the public because it assures the circulation of ideas, creates consensus, and creates the possibilities of choices.

Fig. 5: Our activities

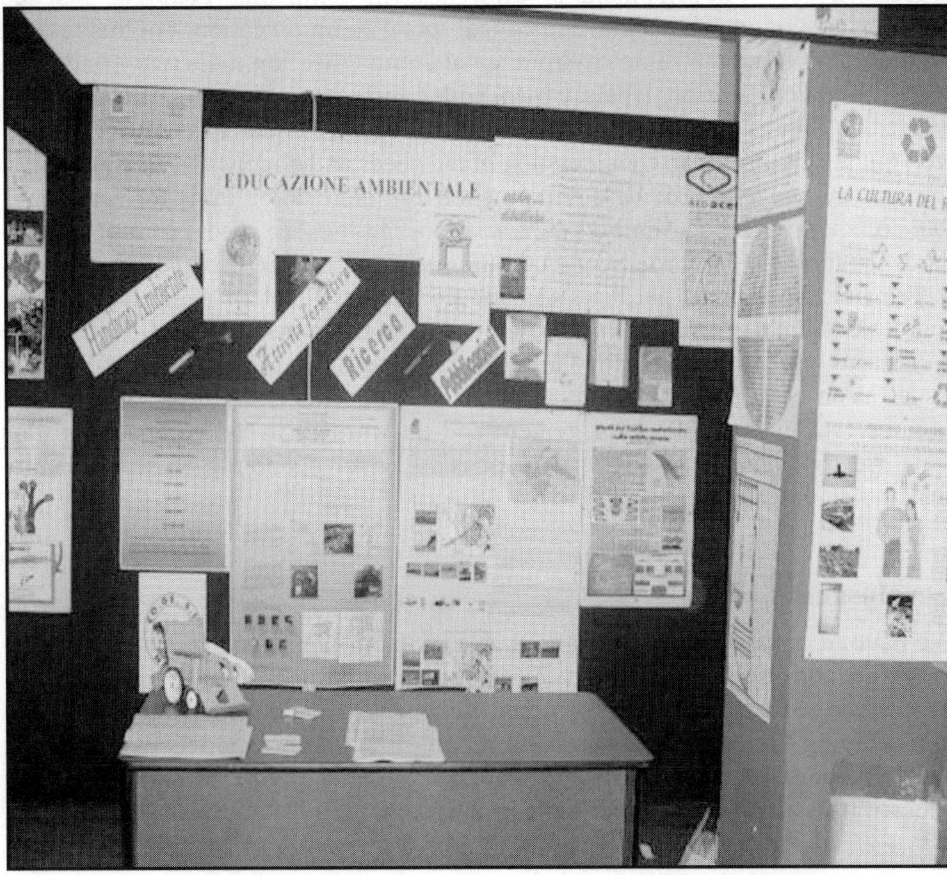


Another activity that we are using in energy saving education is in the context of SICSI, the School of Campania Region that specialises in teaching. In the module of "Ecology", we have extended the traditional teaching topics of ecology and included energy saving which gives new teachers some possibilities of looking at the problems along with their pupils in a critical way. This also gives new teachers the critical capabilities regarding new (also multimedia) tools for education for sustainable development.

## New perspectives for the scientific community for environmental education on climate change

On behalf of the scientific community, for the future we hope to have environmental education activities not only in schools but also in companies. In fact, we trust that, in their responsibilities in terms of air emissions during the productive process, a better development of awareness in companies could be reached through tools of environmental and ethical social communication. For this reason, we are going to create some environmental communication tools (environmental accounting, certification, labels, ethical codes and so on) with the following characteristics. They will not be standardized as European standards (EMAS, Ecolabel etc.) but are calibrated in consideration of the needs and dimensions of companies. In this way, we also provide environmental communication tools for companies that do not have the possibilities (financial possibilities) to access European standards. In this way, environmental communication tools are provided for companies in marginal areas because they are companies that use environmental resources but are at the margins of the main economic and market flows. Finally, we will strongly encourage participation in this process to consider environmental issues as an eco-driver for company development. This is not only for top management but also middle management and all workers in the company. In this way, we can achieve total social responsibility for economic environmentally sustainable development.

## Conclusions

We have to consider climate changes as a problem of risk management because future economic activities will be more and more influenced by human capability to manage the risks caused by climate changes. Today, we have to participate in modifying the energy consumption model through new models of awareness and behaviour. Furthermore, environmental policies could be integrated into educational policies on a wider scale in the region.

## References

### Books

*Aloj, E.* (2000), Educazione Ambientale, Edizioni Grasso, Napoli.

– (2005), Dall'educazione ambientale alla comunicazione ambientale ed etico sociale di impresa, Loffredo Editore, Napoli.

*Cassedy, E.* (2000), Sustainable Energy, a critical assessment, Cambridge University Press.

*Eliasson, B.* (2003), Integrated assessment of sustainable energy systems in China, Springer.

*Leal Filho, W. & Littledyke, M.* (eds.) (2004), International Perspectives in Environmental Education, Peter Lang, Frankfurt am Main.

*Tester, J.W.* (2005), Sustainable Energy: Choosing Among Options, Mit Press.

**Chapter in edited books**

*Aloj, E., De Castro, M., De Iulio, R. & Zollo, A.*: The definition of environmental complexity concept as didactic instrument for sustainability in Atti della Conferenza International Launch in Higher education – Committing Universities to Sustainable Development, 20–23 April, Graz, Austria, 2005, pp. 500–514.

*Aloj, E., De Castro, M., Della Corte, F., Pepe, V. & Zollo, A.*: Strategic activity of Environmental Communication as a didactic goal in the courses of Environmental Education, in Atti della Conferenza International Launch in Higher education – Committing Universities to Sustainable Development, 20–23 April, Graz, Austria, 2005, pp. 500–514.

*Aloj, E., De Castro, M. & Zollo, A.*: The role of the scientific community for the actuation of sustainable development strategies in the connection between the different stakeholders in Walter Leal Filho (a cura di), Innovation, education and Communication for Sustainable development, Peter Lang Editions, 2006, pp. 21–39.

www.arpacampania.it

www.ascuolaconenergia.it;

www.coldiretti.it

www.enea.it

www.energie-rinnovabili.it

Chapter 8

# Teaching Global Climate Change as a Controversial Issue – Active Learning in Higher Education

**Evangelos I. Manolas**

## Abstract

Global climate change is one of the most important controversial issues of our time. If the role of Universities is to create the leaders of tomorrow, then, the duty of teachers in institutions of higher education is to find ways to help students become aware of the problem and understand the complexity of the issue. Following presentation of the definition of controversial issues, the reasons for teaching controversial topics as well as the characteristics of controversial topics, this paper puts forward certain active learning approaches which are in conformity with such considerations without, at the same time, loosing sight of the interdisciplinary nature of the subject. The paper concludes with a discussion of the obstacles which should be overcome so that the methods put forward in this paper can be applied successfully.

## Introduction

Global climate change is one of the most important controversial issues of our time. Uncontroversial scientific evidence that could end the debate is, however, unlikely to become available soon. The intensity of the debate forces us to decide how to act long before conclusive evidence is collected and the climate system is completely understood. Only an environmentally literate citizenry will be able to constructively participate in the on-going discussions (Bord, Fisher & O'Connor 1998; Gayford 2002; McLean 2004; Kohut 2006).

If the role of Universities is to create the leaders of tomorrow, then, the duty of teachers in institutions of higher education is to find ways to help students become aware of the problem and understand the complexity of the issue.

Educating about global climate change is a huge task and an inspiring challenge. Because global climate change is strongly interdisciplinary in character, a wide variety of learning approaches can be brought to bear on the issue. In considering problems related to global climate change one could consider them from several perspectives, including (among many others) economic, political, historical or scientific perspectives. Each perspective has its strengths, and opens up understanding unavailable from other approaches („Introduction: Critical Thinking in Environmental Science" 2007).

Following a presentation of the definition of controversial issues, the reasons for teaching controversial topics as well as the characteristics of controversial topics, this paper puts forward certain active learning approaches which are in conformity with such considerations without, at the same time, loosing sight of the interdisciplinary nature of the subject. The paper concludes with a discussion of the obstacles which should be overcome so that the methods put forward in this paper can be applied successfully.

## Definition of Controversial Issues

Controversial issues can be described as those areas where the data itself, the collection of data, the empirical evidence or the ethics involved are open to various interpretations and viewpoints (Van Rooy 1993).

## Why Teach Controversial Issues?

- they are topical
- they are important
- they are interesting
- they can be related to students' lives
- the teaching of controversial issues provides a chance for staff and students to reflect, to develop, to practice, to comprehend, to evaluate
- the teaching of controversial issues can assist students to develop reasoning skills

## Characteristics of Controversial Issues

A controversial issue is one in which

- there are competing values and interests
- people disagree strongly about statements, assertions or actions
- there is political sensitivity
- emotions become strongly aroused
- protest is aroused

Controversial issues can be about

- what has happened
- the causes of the present situation
- the desirable ends to work towards
- the appropriate course of action to be taken
- the likely effect of action

Controversy

- is usually complex
- requires knowledge, awareness of self and sense of identity
- requires balance, disclosure of commitment and neutrality, i.e. taking a stance without coercion or indoctrination of others („Teaching Controversial Issues", 2007)

## Some Teaching Techniques

There are many ways to teach controversial issues. Each method has its appropriate application depending on the nature of the student population and the type of educational outcome to be fostered. Ultimately, each teacher must decide which learning technique to use and the relative amount of total in-class and out-of-class time devoted to complete the process (Kagan 1992). Some examples of techniques that can be used are as follows:

### 1. Carousel Brainstorming

In carousel brainstorming students rotate around the classroom in small groups, stopping at various stations for a designated amount of time. While at each station, students activate their prior knowledge of a subject through conversation with peers. Ideas shared will be posted at each station for all groups to read. Through movement and conversation, prior knowledge is activated, providing scaffolding for new information to be learned („Carousel Brainstorming" 2002).

The students are first divided into small groups to make role selections. Pre-planning of these groups will most likely ensure a cohesive group with mixed abilities. The teacher highlights the responsibility of the group as „all for one and one for all".

The teacher generates questions that spark interest and conversation for the topic of study. On the subject of climate change such questions could be: What is climate change? What causes climate change? Why are some people concerned about climate change and who are they? Why are some people not concerned about climate change and who are they? What is being done about climate change? Each of the questions is written on a separate sheet of chart paper. The questions are posted around the classroom in stations. Each group's recorder is given a different color marker. The groups are directed to a starting station and given a time limit. Students, then, brainstorm ideas and the recorder for each group write them directly on the chart paper. At timed intervals, groups rotate or „carousel" to each station adding new responses to the list of answers, but may star (*) those responses with which they already agree.

The teacher should maximize the effectiveness of learning by maintaining his / her presence in the classroom. The teacher should be available to students for any needed clarifications as they work. Equalize the rotation time for the groups with a timer and have a pre-planned time signal. Recognize that groups may need additional time as the lesson progresses to read information posted by prior recorders. The instructor should also vary the amount of time necessary according to student needs and rotate groups clockwise to complete the rotation.

At the end students return to their original question and write a summary answer to the question utilizing the responses. Each group then reports to the class the question and answer and a general class discussion of the topic concludes the assessment (Trygestad 2005; „Cooperative Learning with Carousel Brainstorming", 2007).

## 2. Discussion using Cartoons

Cartoons are a powerful means of providing social and political commentary. They may or may not be humorous. Cartoons can be designed to show all the vices or virtues associated with a particular character or event. They can be especially effective in calling attention to the ironies that surround our lives. They may also be biting in their comment on a social issue or in reference to a particular person („Using cartoons to teach social studies", 1999). They usually use the techniques of symbolism and exaggeration to present forcefully particular points of view (Doring 2000).

Cartoons may constitute suitable means for introducing the concept of the greenhouse effect, for reflecting and thinking about the issue of the greenhouse effect, for understanding a point of view, for developing skills in interpreting cartoons. Consider, for example, the following cartoon („The greenhouse effect", 2002):

Fig. 1: Cartoon depicting the greenhouse effect

After an analysis of the symbolism and exaggeration in the cartoon as well as linking of the cartoon with different scientific views on the greenhouse effect students are asked to discuss the following questions: a) What are your emotions about the cartoon? Justify your answer. b) What do you think about the message offered in the cartoon? Do you agree or disagree? c) When was more satisfying looking at the cartoon as a piece of art: before or after the interpretation of it?

Why? d) If you were asked to change the cartoon, what part(s) would you change? Why? What parties would be opposed to the message communicated by the cartoon? Why? (Manolas & Filho 2004)

### 3. Reaction Sheet

After a lecture presenting both sides of the climate change debate, the teacher passes around several sheets to collect written reactions to these three questions: „What ideas do you question," „What ideas are new to you," and „What ideas really hit home?" The answers of the students are followed up with discussion. Variations are to ask each student to write their own sheet or to have small groups do so (Berquist & Phillips 1975).

### 4. Guest Lectures

Guest lectures by people with specialized knowledge and experience on issues related to climate change provide great learning experiences. A lecture by a representative of the Ministry of the Environment or an official from an NGO such as Greenpeace will add variety, that is, the students get to hear from someone other than the professor. The key to making guest lectures work as an effective teaching tool is the question and answer session that follows the guest's remarks. The instructor has a vital role to play as a facilitator, not unlike the chair of a paper session at professional meetings (Donnermeyer 2005).

### 5. Debate

Students are divided into small groups and asked to collect information on the greenhouse effect and evidence for and against the possibility this phenomenon exists. As they collect their information, students should also be identifying what effect global warming might have on the earth, and what possible solutions might exist to any greenhouse effect.

Questions that the groups should consider as they collect information include: What impact might $CO_2$ emissions have on overall global warming? Is there evidence of a strong connection between $CO_2$ emissions and global warming? What impact do chlorofluorocarbons (CFC) emissions have on global warming? How can we model future changes in the global climate with the scientific models available? What can global climate models tell us about life on earth for future generations? How much scientific weight should we place on the results of these models?

The groups gather to participate in a debate about the scientific merits of the greenhouse effect. As participants offer their views, they should support their position with evidence drawn from their research. Positions and evidence should be scrutinized by the group according to scientific merits, verifiability, weight of evidence, and influences by outside parties.

As the discussion continues, students write down what criteria would point to the existence of the greenhouse effect.

Finally, students discuss what effect the greenhouse effect might have for life on Earth („The Greenhouse Effect Debate", 2007).

## 6. Essay Writing

Essay writing is among the most popular active learning techniques used in Universities. One idea is to ask students to write an essay debating both sides of the issue, e.g. What is the greenhouse effect? What is its importance to global climate? (Botkin & Keller 2003)

Another idea is for students to write an essay about global warming from the viewpoint of a different person.

Students can choose to write an essay as an environmentalist who is a believer in global warming. The essay can be about environmental policies and ideas to stop global warming, and why they will be effective.

Or they can write as an industrialist who is sceptical about global warming. This essay can be about new environmental laws to stop global warming. They can write about how they will affect his / her business, and what he / she thinks should be done instead.

Whether they choose the first or the second activity it should be made clear to students that they should support their cases with the strongest facts they can find. („Global warming" 2007)

## 7. Research

Students need to be given opportunities to engage in the kind of research that their teachers practice. Such opportunities will bring them into contact with the complexities and confusions of the research world and will offer them the chance to experience the joy and pride of discovery. Teachers, therefore, should involve students in research projects in which they are required to develop hypotheses and gather and analyze data (Neal 2003).

Do adults believe that global climate change is occurring? What are adults doing about climate change? Do young people have similar views? Students may be asked to design and carry out research with the task of identifying and analyzing attitudes and behaviors of people about climate change (Trygestad 2005). At the end, students could present their research in a class research symposium similar to what is done at professional meetings. Such research symposia should be attended by other faculty members and students.

Students who for various reasons do not want to participate in such research could engage in alternative activities. For example, they could locate similar research which has been carried out using the same method and the same or similar questions, summarize and present the results of these works and discuss the advantages and disadvantages of the research method used. By having students conduct research and locate and discuss other and related research, the traditional roles of student and teacher can become blurred as students and teachers share and learn current developments (Houser & Vaughan 1995) about an important environmental issue.

### 8. Student Self-Evaluation

Following completion of their essay (or project) the instructor can ask the students to write a brief evaluation of their learning. For example, they can be asked to answer the following questions:

- What problems did you face during the writing of this essay?
- What solutions did you find for those problems?
- What do you think are the strengths of this essay (project)?
- What alternative plans for this essay (project) did you consider? Why did you reject them?
- Imagine you had more time to write this essay (work on this project). What would you do if you were to continue working on it? (Middendorf & Kalish 1996)

## What are the Barriers?

While the activities discussed in this paper might seem appealing, they often seem appealing for other instructors or other subjects – but not for our own. Trying these activities might seem like inviting a disaster, especially when it means giving up the control that a lecturer commands. In the teaching strategies put forward here the teacher's role is more in the background, where he / she may observe, listen and assist students only as needed. Instead of giving answers to students, the instructor will often set back and watch the students struggle through the problems. The teacher also takes the role of questioner, asking members of the class about their conclusions or solutions to problems, asking them to justify what they did and why.

The instructor may be discouraged by students who resist an activity that appears challenging and difficult, forces them to think, and does not allow them to be passive learners, because students are used to sitting in lectures where they are not required to talk, solve problems or struggle with learning new material. Students may want the teacher to do more explaining, and giving them answers rather than struggle with a problem themselves. Sometimes this concern is related to the issue of grading fairness. Students may feel it is unfair to give one grade to the entire group rather than separate grades to individual students, especially if students do not contribute equally. By adopting a grading policy where individual student contributions are taken into account may rectify the situation (Garfield 1993).

Other obstacles associated with the teaching methods suggested include: limited class time, a possible increase in preparation time, the potential difficulty of using such activities in large classes, the fear of teaching in unorthodox ways and taking risks in the classroom, the fear of not covering as much material as covered in lectures, the fear that the time spent for organizing and implementing such active learning activities would take time away from research and writing (Bonwell & Eison 1991; „Active learning: Getting students to work and think in the classroom", 1993).

Each of the above obstacles, however, can be overcome through careful, thoughtful planning and through support from University authorities. The way we instruct can have profound implications for the way students behave towards the environment. Institutions of higher education have a responsibility to themselves, their students and to society, to guide individuals and prepare them for the challenges of our environmental future.

## Conclusion

This paper discussed ways in which global climate change as a controversial issue can be effectively taught in institutions of higher education. The instructional strategies suggested were carousel brainstorming, discussion through the use of cartoons, reaction sheets, guest lectures, debate, essays, involving students in research projects and student self-evaluation. The use of such methods can help students acquire skills such as listening, handling conflict, self awareness, distinguishing between fact and opinion, critical thinking, ethical and logical reasoning, recognizing bias or evaluating evidence. Equally as important, students develop empathy and a willingness to perceive and understand the interests, beliefs, and viewpoints of others. These are critical skills not only for tackling important environmental issues but also for decision making and social action in an increasingly complex world.

## References

„Active learning: Getting students to work and think in the classroom" (1993), Speaking of Teaching, vol. 5, no. 1. Available www.ntlf.com/html/lib/faq/ac-stanford.htm

*Berquist*, *W.H.* & *Phillips*, *S.R.* (1975), „Classroom structures which encourage student participation". In Quel, G.H. (ed.), A Handbook for Faculty Development, The Council for the Advancement of Small Colleges in association with the College Center of the Finger Lakes, pp. 118–121.

*Bonwell*, *C.C.* & *Eison*, *J.A.* (1991), „Active learning: Creating Excitement in the classroom". Available www.ntlf.com/html/lib/bib/91–9dig.htm

*Bord*, *R.J.*, *Fisher*, *A.* & *O'Connor*, *R.E.* (1998), „Public perceptions of global warming: United States and international perspectives", Climate Research, vol. 11, pp. 75–84.

*Botkin*, *D.B.* & *Keller*, *E.A.* (2003), Environmental Science. Earth as a Living Planet, John Wiley & Sons, Inc., New Jersey.

„Carousel brainstorming" (2007), Available http:// its.guilford.k12.nc.us/act/strategies/carousel_brainstorming.htm

„Cooperative learning with carousel brainstorming" (2007), Available www.glencoe.com/sec/teachingtoday/weeklytips.phtml/print/223

*Donnermeyer*, *J.F.* (2005), „Making Introductory Rural Sociology Interesting to a Non-Interested Audience", NACTA Journal. Available www.findarticles.com/p/articles/mi_qa4062/is_200509/ai_n15614175/pg_10

*Doring*, *A.* (2000), „The Use of Cartoons as a Teaching and Learning Strategy". Available www.tedi.uq.edu.au/Conferences/teach_conference00/papers/doring.html

*Garfield*, *J.* (1993), „Teaching Statistics Using Small-Group Cooperative Learning", Journal of Statistics Education, vol. 1, no. 1. Available www.amstat.org/publications/jse/v1n1/garfield.html

*Gayford*, *C.* (2002), „Controversial environmental issues: a case study for the professional development of science teachers", International Journal of Science Education, vol. 24, no. 11, pp. 1191–1200.

„Global warming" (2007). Available http:// weathereye.kgan.com/expert/warming/assign.html

*Houser*, *N.* & *Vaughan*, *S.* (1995), „Meaning in Transaction: A Socio-Psychological Reconsideration of Teaching and Learning", Curriculum and Teaching, vol. 10, no. 1, pp. 17–28.

„Introduction: Critical Thinking in Environmental Science" (2007). Available at: http:// abacus.bates.edu/~raustin//FluxFlow/L1.3%20critical%20thinking.pdf

*Kagan*, *S.* (1992), *Cooperative Learning*, Resources for Teachers, Inc., San Juan Capistrano.

*Kohut*, *A.* (2006), „Little consensus on global warming", The Pew Research Center For The People & The Press, Washington, D.C. Available http:// people-press.org/reports/print.php3?PageID=1062

*Manolas, E.I. & Leal Filho, W.* (2004), „The Use of Cartoons in Environmental Education: A Case Study of a Learning Approach," Discursos, Special Issue, pp. 399–405.

*McLean, J.* (2004), „The disputed science of global warming". Available http:// meteo.lcd.lu/ globalwarming/ McLean/ Disputed_Science_of_Global_ Warming.pdf

*Middendorf, J. & Kalish, A.* (1996), „The 'Change-up' in lectures", TRC Newsletter, vol. 8, No.1. Available www.indiana.edu/~teaching/allabout/pubs/changeups.shtml

*Neal, E.* (2003), „Active learning beyond the classroom". Available www.ucet.ufl.edu/essays/topic1–4.htm

„Teaching controversial issues" (2007), Available at: www.finders.edu.au/teach/t41/inclusive/controversial.php

„The Greenhouse Effect" (2002), Tata Energy Research Institute, Available http:// edugreen.teri.res.in/explore/climate/greenhs.htm

„The greenhouse effect debate" (2007), Available at: http:// commtechlab.msu.edu/ sites/ letsnet/ noframes/ subjects/ science/ b5u112. html

*Trygestad, J.* (2005), „Climate change: The future is upon us". Available www.wwu.edu/ depts/ castudies/ k12studycanada/ files/ lesson_plans/ 2005/ J%20Trygestad%20Climate%20Change.pdf

„Using Cartoons to Teach Social Studies" (1999). Available at: www.homepages.dsu.edu/ smithju/ archive/ Fall1999/ seed302/ not%20used. cartoons.htm

*Van Rooy, W.* (1993), „Teaching controversial issues in the secondary school science classroom", Research in Science Education, vol. 23, no. 1, pp. 317–326.

## Chapter 9

# Regional Climate Change and Coasts
## –
## A Case Study on Perception, Information, Dissemination and Education

**Gerald Schernewski, Steffen Bock, Holger Janßen, Nardine Löser, Annemone Hmielorz**

### Abstract

The Oder/Odra estuary region serves as an example to describe the concrete regional state-of-the-art with respect to Climate Change. We evaluate the present knowledge and awareness among locals and authorities and introduce ongoing information, communication and dissemination measures, campaigns, and pathways. We try to answer questions about the efficiency of existing information and communication measures from a regional perspective, outline short-comings, and discuss some future perspectives.

The wide range of potential Climate Change effects on our coast and our coastal ecosystems as well as their social and economic consequences are not well perceived. One reason is the lack of concrete knowledge. Regional long-term forecasts as well as reliable regional scientific results are increasing, but still not abundant. Further, the transfer of knowledge between scientists, regional practitioners and locals is not efficient. We further learn that Climate Change requires an improved information, communication and education strategy, involving different media and dissemination path ways. To measure the efficiency of information offers and campaign remains a major task for the future. However, personal contacts and communication are still one of the most important pathways to transfer information.

### 1. Background and objectives

The problem of Climate Change and its consequences are nowadays well present in all kinds of media, are communicated to the broad public and are widely and intensively discussed. For Germany, we can say that there is a general awareness about this issue. As soon as we look at a specific region and its concrete challenges due to Climate Change the picture becomes more complicated and a more detailed analysis is necessary. We focus on the Oder estuary region and are going to tackle the questions: (a) Are people and decision makers informed about the consequences in their specific region and is there awareness about the entire spectrum of possible Climate Change consequences? (b) Are the regional media a suitable source for information and is this information balanced and of sufficient quality? (c)What are possible additional tools and measures to communicate regional Climate Change consequences and to raise awareness? (d)What are the possibilities and limits of the Internet in this respect? (e) How efficient are these communication and information tools and (f) Does the knowledge about Climate Change already show consequences? Are there regional strategies to deal with it and if, how can these processes be supported?

Fig. 1: The Oder/Odra estuary region

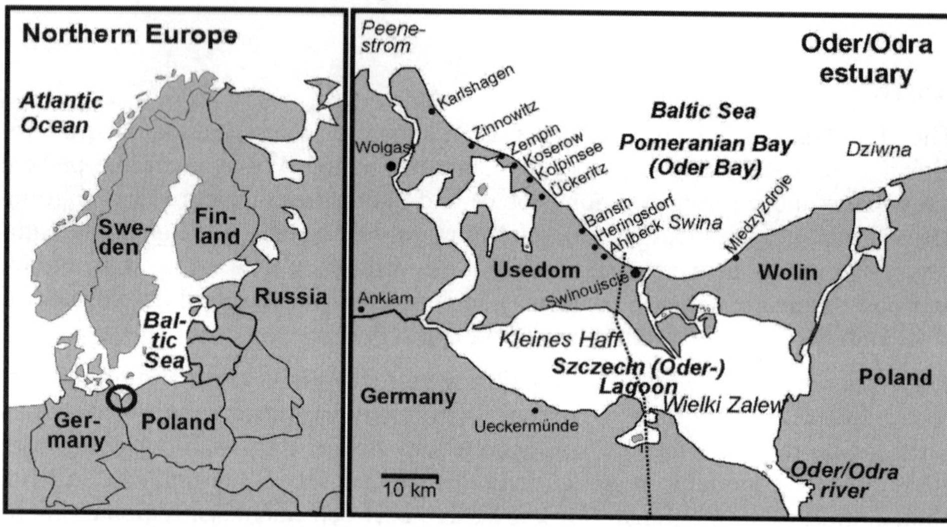


## 2. A changing coastal zone under pressure

### 2.1 The Oder/Odra estuary

The Oder river (Polish: Odra) (854 km length) is one of the most important transboundary rivers in the Baltic region (Figure 1). Its basin (118,000 km²) is shared between Poland (89%), the Czech Republic (6%) and Germany (5%). It is a lowland river with its origin in the hilly Polish-Czech boundary region. Due to the warm-temperate climate with sufficiently rain in all seasons, the water discharge shows only a limited annual cycle with an average discharge of 530 $m^3/s$. Floods are a rare phenomenon and are not necessarily linked to snow melting. The last extreme flood with a discharge up to 2,800 $m^3/s$ took place during late summer 1997 and caused severe damage.

The Oder estuary is a complex pattern of lagoons and islands and shared between Germany and Poland. The Oder river first enters the large (680 km²), shallow Szczecin (Oder) Lagoon. The river discharge and its nutrient and pollutant loads are responsible for the lagoon's highly eutrophic state. The lagoon serves as a storage pond for sediment, nutrients and heavy metals and helps to protect the Baltic Sea from pollution. The water flushing time is only 55 days and the lake-like salinity around 1.5 $^{0}/_{00}$ shows that the lagoon is only to a minor degree influenced by the Baltic Sea. The lagoon water is discharged into the Pomeranian (Oder) Bay (Baltic Sea) via three outlets.

The coastal landscape is flat and dominated by agricultural land and forests. In some areas sand, gravel, oil and gas are exploited. Broad reed belts and artificial sandy beaches near the few small towns characterize the coastline. Due to its outstanding ecological value and beauty, most of the coastal area is under nature protection. With about 840,000 inhabitants (414.000 in the city of Szczecin) the estuary region is only sparsely populated.

Tourism, agriculture, fishing (3000 t/a in the lagoon) and shipping are important economic activities (Figure 2). Details about the river basin and the coast are given in Löser & Sekścińska (2005) and Behrendt & Dannowski (2005).

Fig. 2: Impressions from the Oder lagoon (left) and the Baltic Sea coast (right)

## 2.2 Regional transformation processes

The Oder estuary faces dramatic political, social, economic and natural changes. This is important because these changes are overlapping with Climate Change and conceal the effects.

*Political and social changes:* The German part of the Odra region belonged to the former socialistic German Democratic Republic (GDR). With the German re-unification in 1990 important social, political and economic changes took place and are still ongoing. Despite huge financial efforts the Odra region fell behind the development of other parts of Germany. Like eastern Germany, Poland was subject to social changes and its transitional economy is still facing serious ongoing problems and changes. During the last decade the economic and social developments in Germany and Poland were largely independent and caused strong social and economic gradients. In the coastal area social problems are increasing and the gap between the flourishing seaside resorts and the hinterland is still deepening.

*EU-membership and transformations:* With Poland's EU-membership in 2004, the entire Odra region became part of the European Union. The new agriculture and industry policy, but also the implementation of new standards will cause dramatic changes. Cross-border cooperation as well as competition will increase and cause social and economic transformations. This will have multiple effects on the Odra basin, the river and the coastal area.

*Legal challenges:* The active European environmental policy has led to the Marine Strategy, the recommendations on Integrated Coastal Zone Management, the Habitat Directive (Natura 2000) and the Water Framework Directive (WFD). Especially the WFD and Natura 2000 are now being implemented in Germany and Poland. Natura 2000 will lead to a large number and a network of protected areas in the Odra region which require a management. Even more important is the WFD which has the aim to ensure a good water quality in all member states. The WFD is a cross-border and river basin – coast approach and requires many measures to improve water quality during the next years.

### 2.3 Regional Climate Change effects

Climate Change will have multiple effects on the Baltic Sea (e.g. Schernewski & Schiewer 2002; Schernewski & Löser 2004) and on the region. Some important effects shall be briefly mentioned to show the complexity and interdependencies. We will restrict ourselves to temperature, precipitation and sea-level rise and focus only on the estuary and the shoreline.

*Temperature:* Higher water temperatures in winter will reduce the spatial coverage, thickness and duration of ice. Lacking ice improves the light penetration and availability in the water body. This has consequences on many biological processes. Algae developments and „blooms" e.g. can take place earlier and alter the entire annual biological succession as well as the food-web.

Higher summer temperatures may be linked to more stable, calm and sunny periods. This would increase the likelihood of vertical temperature stratifications in the water body and of internal eutrophication. Internal eutrophication happens when a stratification hampers vertical mixing and an oxygen transport towards deeper water layers. Degradation processes deplete the available oxygen and the water as well as the sediment become anoxic. In the lagoon, large amounts of phosphorus are bound to iron and these compounds become soluble under anoxic conditions. As a result, large amounts of phosphorus enter the water body and increase eutrophication. This process already has been observed during hot summers.

Nearly all bio-geo-chemical processes in the water are temperature dependent. Higher water temperatures decrease the solubility of important gases like oxygen or carbon-dioxide. They increase denitrification, an important sink-process, which transfers soluble nitrogen compounds into molecular nitrogen. Molecular nitrogen is lost to the atmosphere and reduces the availability of the nutrient nitrogen in the water.

Already today, many microbial, phytoplankton, zooplankton, benthos and fish species in our aquatic ecosystems are alien species, which replace our native species. Increasing temperatures allow new species to settle and to survive. The species composition and the functioning of our ecosystem will change. Some new species might become a plague, like jellyfish along the southern Baltic coast in 2006.

But even our native species will be affected by higher temperatures. The species composition, abundance and annual succession depend on the physical and chemical environment, which is effected by Climate Change. The occurrence and intensity of algae blooms will change. Even the size of species might change. Some zooplankton species e.g. become smaller when the temperatures are higher, because they need more energy for staying in the water body and to prevent sedimentation.

Hygienic water quality (bathing water quality) is effected by higher temperature as well, because coliform bacteria and human-pathogen viruses survive longer and are spread further. (Schernewski et al. 2001). On the other hand higher water temperatures are beneficial for bathing tourism, as well.

*Precipitation:* The amount and annual pattern of precipitation in the river basin largely determines the Oder river water discharge as well as the loads of nutrients and pollutants entering the coastal waters (Wielgat & Schernewski 2002). Recent climate change scenarios do not assume significant changes in the annual amount of precipitation, but the temporal pattern might change. Wet winters and dryer summers have a higher likelihood in future. The recent years 2003 and 2005 already show possible consequences for the coastal waters. In dry summers the water discharge and the nutrient load in the river is reduced. Usually nutrients are abundant in the lagoon. It seems that in dry years, the primary production in the lagoon can be much more limited by the lack of nutrients, especially nitrogen. This hampers strong algae blooms, but favours the development of blue-green algae. Many of these species are at potentially toxic. These nitrogen limitation periods may coincide with internal eutrophication and the release of phosphorus from the sediment. As a result large amounts of phosphorus are discharged into the Baltic Sea and might cause increase algae developments, because algae in the coastal Baltic Sea are often limited by the availability of phosphorus during that time.

Higher precipitation in winter and more extreme weather events may increase the likelihood of river floods. The Oder flood in 1997 clearly showed the consequences, damages and costs. An outstanding situation might occur, if a river flood coincides with storm surge. The coast would be in squeeze and be endangered from both sides. Outstanding storm surges like in 1995 or 2002 can reach a level of 1.7 m in the Pomeranian Bay. However, it is uncertain if the risk of storm surges will increase in future.

*Sea level:* The Pomeranian coast sinks at a rate of 1 mm/a. Therefore, until 2100, a relative sea level rise of 10 cm will take place. The sea level rise due to climate change can be up to 88 cm until 2100 (IPCC 2001). This will cause increased coastal erosion, higher coast for protection measures and endanger tourism infrastructure (sport boat harbours, beaches, piers, promenades). This sea level rise increases the effect of storm surges.

Usually, warm winters are a result of strong westerly winds, which transport warmer Atlantic air-masses towards the Baltic Region. These strong westerly winds increase the sea-level along the Baltic coast in Germany. In the warm and windy January and February 2007, these high Baltic Sea levels caused an ongoing pressure on dikes. The dikes were soaked through and, especially in the lagoons and estuaries, became instable. This unexpected threat is possibly more important then storm surges in the Baltic Sea and might become a major problem in future.

## 3. Regional knowledge about Climate Change: A newspaper analysis

An estuarine system, like the Oder region, is extremely vulnerable to Climate Change. It faces the sea on one side and has the river on the other and both will be strongly affected by Climate Change. Further, the region is dominated by lowlands and the economy largely depends on summer tourism. Damages to the infrastructure, the beaches or the coastal environment can cause a fast decline of tourism and endanger the economy of the region. Therefore, the public in the Oder estuary region should be well aware of the possible dangers and problems due to Climate Change and one could expect that strategies and plans already exist to deal with the threat.

Is this true? Is there a regional public knowledge and awareness about Climate Change and its possible consequences? To answer this question one usually would develop a questionnaire and carry out a survey with a representative number of people (e.g. Peters & Heinrichs 2005). This is a laborious and time-consuming task and could not be carried out within the projects IKZM-Oder and ASTRA. Therefore, we had to look for alternatives. We systematically analysed the freely available online Internet-archives of the major regional newspapers on the German (Ostsee-Zeitung) and Polish (Kurier Szczecinski) side of the border (Hmielorz & Löser 2006). Our hypothesis is that these newspapers are the major source of regional information for the public. We assume that only if Climate Change and its consequences are well reflected in the newspapers, a public awareness in the population can be expected.

On the German side, the regions Greifswald and Usedom/Peene (the major parts of the Oder estuary) have a total population of 143.300 and about 29.000 daily issues of the Ostsee-Zeitung are sold. 39% of the 26,223 households subscribed the Ostsee-Zeitung. We can say that it is the predominant print product and cover the region spatially well. The same hold for the Polish Kurier Szczecinski. Not all articles are entered into the online-archives. Only about 75% (20%) of the Ostsee-Zeitung (Kurier Szczecinski) articles are available online. Clear explicit criteria for the choice of articles to go online do not exist. Our analysis covered the period between January 2000 and May 2006 and was restricted to articles dealing with regional aspects. 15 keywords with relevance to climate were attributed to five categories: (a) heat and draught; (b) flood, flood protection and storm surge; (c) Climate Change; (d) weather and (e) others (like blue-green algae, Climate and carbon-dioxide and nature protection) (Figure 3). The articles were analysed with respect to content and length. Additionally, five interviews with regional journalists were carried out.

Altogether 175 regional and climate relevant articles were found online. 138 published in the Ostsee-Zeitung and 37 in the Kurier Szczecinski. During the six analysed years, the number of articles with a regional focus increased. Nowadays, we can assume that climate relevant articles are about once a week in the newspapers. The number of articles about tourism, which is the major economic sector and predominant source of income, is 46% (Kurier Szczecinski) respectively 71% (Ostsee-Zeitung) higher. Climate is important for many economic branches (e.g. agriculture, fisheries and tourism). Therefore, climate is an important regional topic and well reflected in the newspapers. But articles about climate do not necessarily deal with Climate Change.

In the German Ostsee-Zeitung, the topics flood, flood protection and storm surge as well as weather are prevailing. In the Polish Kurier Szczecinski, heat and draught as well as weather is content of most articles. Both newspapers focus on short- to medium-term weather events. Long-term climate and Climate Change are only rare topics. Therefore, we cannot assume that the population is well aware and informed about regional Climate Change and its potential consequences.

Fig. 3: Categories used in the online newspaper archive analyis (2001–2006). All climate relevant articles were allocated to one category. For details see the text.

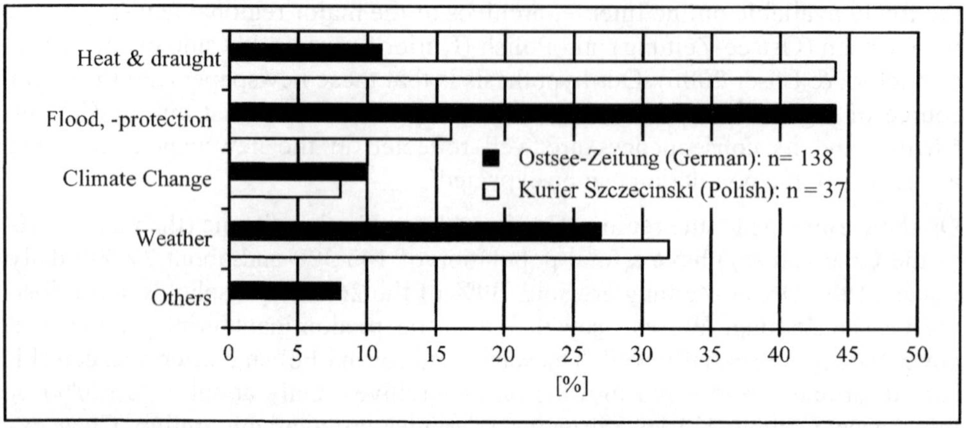


The newspaper analysis is a fast method to get a first impression about the regional importance of Climate Change in the public, but it has limitation and short-comings. It cannot replace detailed surveys. Newspaper online-archives are still under development and growing, only a percentage of the printed article are available online and the criteria for publishing an article online are not well defined. Therefore, temporal analyses are possible only to limited degree. So far, we did not compare print issues and the online articles to get an impression of the online publishing criteria, the total amount of printed relevant articles and possible systematic shifts in the thematic focus. It is likely that articles with long-lasting and supra-regional importance have a higher likely-hood to be published online. Our key-words might not have found all relevant articles. The evaluation whether an article is relevant as well as the classification into categories has been done manually. This implies some subjectivity. Further, we do not know how strong the impact of the newspapers is and how efficient articles with respect to awareness-rising are.

## 4. Internet based information, dissemination and education

How can we overcome information, communication and education deficits with respect to Climate Change? The Oder estuary region is relatively large, divided between Germany and Poland and many stakeholders, projects, authorities, administrations or ministries are directly or indirectly affected by Climate Change. They have their own responsibilities, strategies and position and are active player in this field. In a diverse region like this, it is hardly possible, to develop one joint strategy to overcome all information, communication and education deficits. Instead, a strategy and tools are needed to link, coordinate and harmonise all these activities and to ensure an efficient information flow across the existing hierarchies and stakeholders. Additionally, well designed demand-driven information tools and materials are needed, to fill obvious gaps. Our projects do not have the critical financial and staff mass and the necessary political weight to develop a comprehensive Climate Change information and communication strategy. We had to focus on additional tools and materials to support the information process. We provided regional newsletters, posters, brochures, e-learning modules and an internet-based information-system. In the following chapters we present these tools and critically evaluate their success.

### 4.1 E-newsletters

Newsletters distributed via email are a cheap, flexible and fast possibility to disseminate information. It is easily possible to provide all newsletter issues permanently on a webpage and to ensure access to older issues. The regional newsletter for the Oder estuary has a broad scope on sustainability. Climate Change information is only sometimes a topic. To link Climate Change information to other related topics ensures a critical mass und allows several issues per year. An advantage of this newsletter is that the information can easily be picked up by other media and other newsletters. The national German coastal newsletter e.g. uses these articles on one hand and on the other hand provides information for the region, as well. Disadvantages are that the target groups have to have email and Internet-access and that it takes time until a newsletter (even if it is free) is well accepted and has a sufficient audience. Finally, the flood of available newsletters limits the willingness to accept another one. A systematic evaluation of the acceptance and impact of this regional newsletter s lacking, but has been carried out for the national version. The „Küstennewsletter" exists already for 3 years, has nearly 1000 subscribers and every new issue is downloaded from the internet about 1500 times during the first two weeks after the release. Many subscribers contribute content to the newsletter and an online survey clearly showed that the readers are satisfied. However, the time will show, if a regional newsletter on sustainability including Climate Change can get a critical mass of readers, as well.

### 4.2 E-learning modules

During recent years, the internet platform „IKZM-D Lernen" has been established and extended (www.ikzm-d.de/english.html). It is an online learning system consisting of independent online study, information and teaching modules. The modules deal with the coast and the sea in general and with Integrated Coastal Zone Management (ICZM) in particular. The modules are free of charge and address experts, students and interested citizens. They are largely in German but some modules in English are available, as well. Several modules are about Climate Change or are at least dealing with Climate Change aspects. Similar to newsletters, online modules are a cheap, flexible and fast possibility to prepare and provide information. Authors easily can adapt the module to the need of the audience and have always the possibility to update and modify the content. Due to a simple handling and the provision of comfortable editing tools, the preparation of online modules is attractive for teachers and for students. During the summer-school „Coastal and Marine Management" in 2005, students prepared information modules on coastal and marine uses, problems and perspectives. Climate Change was the topic of one module.

Is the provision of online modules an efficient way to support awareness-rising and education on Climate Change? In 2007 it is likely that about 15.000 visitor will access the Climate Change module. In average, every visitor has a look at two pages. Most visitors are coming from search engines, like Google. It clearly shows that Climate Change is a popular topic, but it seems that many visitors access the online module more or less accidentally and have only a brief look at it. Detailed reading is the exception. Similar modules in German and Polish language are much less visited, but due to less competition they are very high ranked in Google and easy to find. Therefore, modules in national languages can be very useful.

However, our access statistics clearly indicates that freely available online modules are not really used for self-education. Tests are provided online but a certificate for successful students is lacking. This, a general reluctance to use online education and possibly a lacking trust in the quality of the modules are possible reasons for the limited success and attractiveness. The modules are valuable as additional material for well defined and certified „traditional" courses and provide some extra-information. Further, students are usually highly motivated if they have to prepare own modules. Therefore, the preparation of modules on topics, like Climate Change, by students is a successful possibility to raise awareness on the topic. However, what we learn is that online modules have to be linked to „traditional" education systems. Especially modules with a regional focus require a close cooperation with regional organisations to meet their demand and to get an overview about regional interests and short-comings. Further, public relation and advertisement is needed to make the modules known to potential users.

### 4.3 The Regional Informationsystem „IKZM-Oder"

During the project ICZM-Oder the „Information System Oder Estuary Region" (www.ikzm-oder.de) has been established. It contains a wide range of information. Scientific data, general regional information is included as well as all kinds of documents, publication or statistics. The information system links regional information, communication and dissemination tools like the databases, the geographical information system and the decision support system and makes it available to a large community. It addressed scientists, authorities, the local population as well as visitors and tourists.

The information-system was launched in May 2004. The number of visitors increased fast and steadily. At the end of 2006, after 19 month, the monthly number of visitors exceeded 9000 and they open about 200,000 pages (Figure 4). About 300 visitors per day is a surprisingly high number, especially if we take into account that it is still growing.

Fig. 4: The Internet regional Information System „IKZM-Oder": Number of users and accessed pages between 2004 and 2007

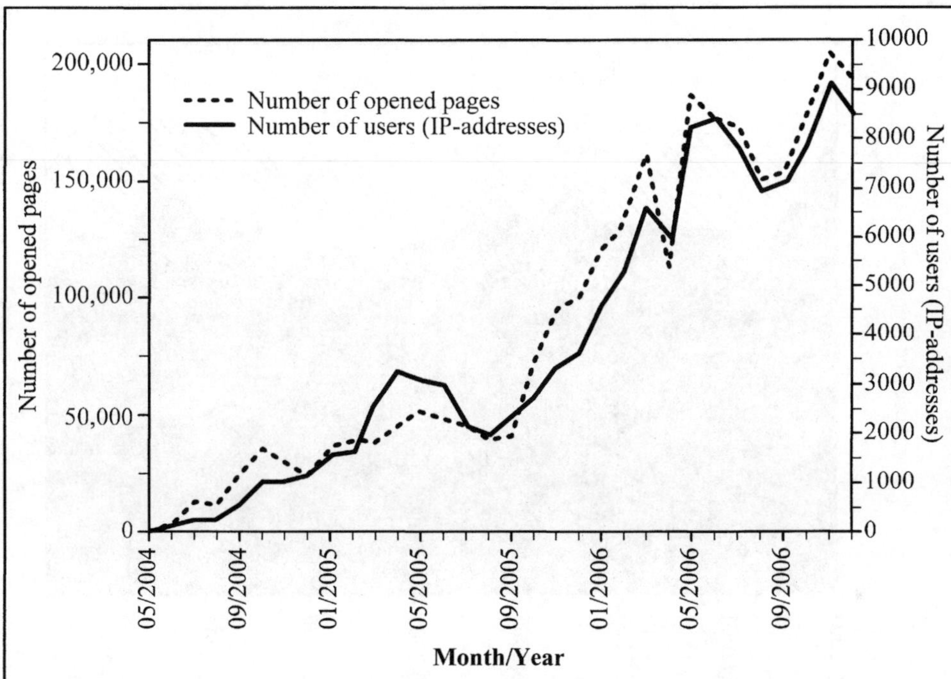


The Oder estuary is a cross border region. The information-system has the task to provide regional information, to increase the knowledge and improve the information exchange on both sides of the border. Therefore, the system is trilingual. Altogether, the system contains 110 html-web pages. Some pages are provided in several languages. At the moment, about 56% of all pages are in German, 22% in English and 22% in Polish. The statistics of the relative access to pages in a certain language reflects this. 22% of the pages are in Polish and they have a share of about 20% of all visited pages. Polish pages are as often accessed as German and English pages. We further can have a look from which country visitors are coming. If we compare only Germany and Poland, we see that 38% are from Poland and 62% from Germany (Figure 5). There seems to be a relatively high interest in the Information system on the Polish side. About 75% of all visitors cannot be allocated to a specific country because their IP-address or provider do not allow it. Therefore, the statistics bears some uncertainties.

However, what we learn is that the information system serves its purpose well and is suitable to serve as a carrier for information on regional Climate Change and its effects in the estuary region.

Fig. 5: The Internet regional Information System „IKZM-Oder": Relative amount of users accessing pages in English, Polish and German between 2004 and 2007

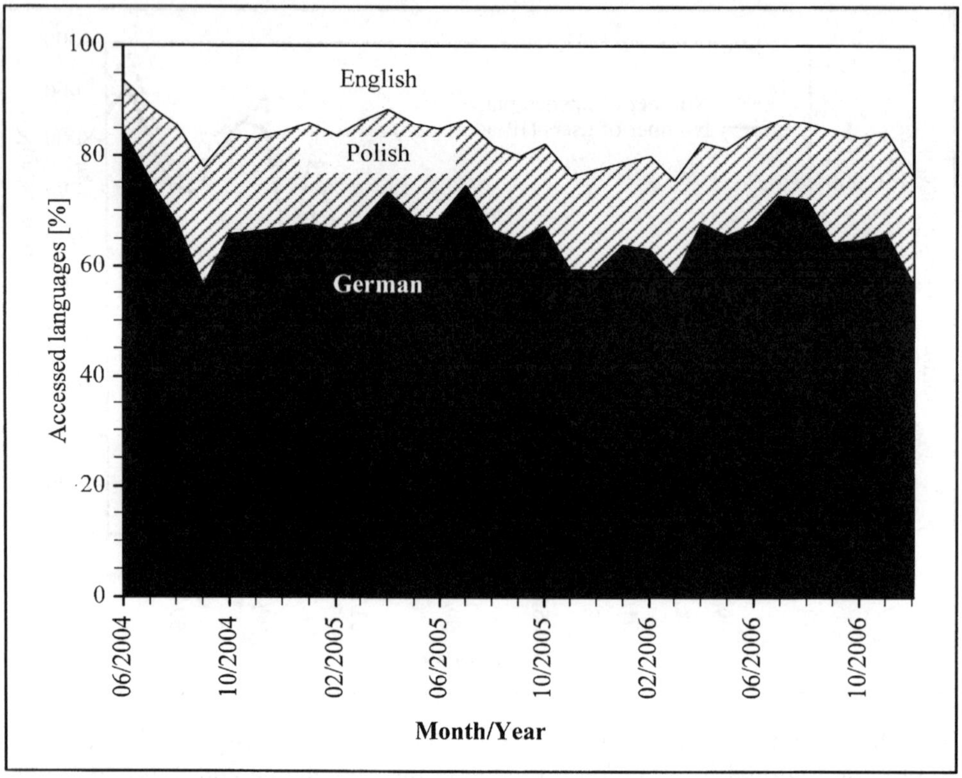


## 4.4 Print products

Regional information tools and dissemination paths have to address and reach a broad but diverse audience. Internet-based tools require activity in getting information and fail to reach interested, but passive people. Therefore, we also prepared supplementary print products like flyers, brochures, posters and on-line information but also glossy magazines for interested laymen (Figure 6).

Only several products, with specific tasks, allow to address a diverse audience. The depth of information, the used language (e.g. basic language), the entertainment factor and the kind of presentation differs among the print products. They have different advantages and disadvantages. E.g. flyers can easily be distributed to a broad public and can be a teaser to inform about information offers. They are only of limited value to transport the information itself. Posters are a comparatively cheap and easy possibility to publish and present information. Posters present information in a coloured and lively style, which can help to impart knowledge about complex issues, while the limited space does not allow comprehensive information. Disadvantage is further that space to hang them out is needed at attractive places and restricts the dispersal.

Fig. 6: Print products and the online portal to information modules

One attempt to bridge the gap between traditional environmental information offers and the addressed audience is the magazine „Meer & Küste" (Sea and Coast). With a specific coastal and maritime view on Baltic issues, traditions and developments this magazine (circulation: 50,000 copies) is attractive to tourists as well as local inhabitants. It is a glossy magazine which tries to inform laymen in a serious but also entertaining way and it is distributed for free. Of course it is not without effort to publish such a magazine but it might be an essential component in a chain of information offers.

All print products require efficient distribution pathways and a good dissemination strategy. For these reasons a co-operation with local authorities, stakeholders and other disseminators and distributors is required. The combination of print products with on-line tools can be a useful strategy to address interested laymen. The transfer of information via several path ways and media as well as over longer times, ensures that people are frequently exposed to this information and helps to develop a permanent knowledge. However, we did not analyse the success and efficiency of our Climate Change information campaign, so far.

## 5. Information, communication and education – Evaluation and perspectives

### 5.1 Climate Change and newspapers: Is the public well informed?

Newspapers are an important, but not the only source of information. Radio and television are possibly of similar importance and, as mentioned, the newspaper analysis contains several methodological short-comings. However, our impression is that the regional public is not well informed and aware of Climate Change and its regional effects. What are the reasons?

The interviews with journalists give some additional insights. Climate Change is a complex topic with still many uncertainties. It is not easy to communicate to a broad pubic and it requires longer articles. Space limitation becomes a problem. The journalists have the impression that Climate Change does not really affect the readers directly and therefore, is of minor interest for them. Usually, articles about climate relevant issues have a negative message, do not show perspectives and solutions and leave arouse but helpless readers. Further, they complain about insufficient information material and sources and wish a closer communication with scientists. Scientists are asked to become more active with respect to information and dissemination of their results.

The Oder estuary region is facing social and economical changes, shows strong gradients and a high rate of unemployment. These economic problems and changes are short to medium-termed and very present in the awareness of the population. They limit the willingness to deal with slow and long-term problems, like Climate Change effects. Even on the policy-making level, the time horizons are legislative periods and hardly much above. Therefore, it is necessary to make the immediate importance obvious to the population.

Our projects ICZM-Oder and ASTRA are dealing with Climate Change effects and generated a lot of contacts to different regional authorities and key-players. Our impression is that on the authority level the awareness about Climate Change is well developed. Especially, the consequences of sea-level rise have already caused modifications of planning and measures are being implemented. However, the authorities work largely without interaction and are focused on their specific needs and responsibilities. An integrated, trans-disciplinary approach or planning, across the hierarchy levels does not exist, so far.

Sea-level rise is by far the best known consequence of Climate Change. This is probably true for all regional stakeholders as well as the local population. Measures towards an improved storm surge protection are already taking place. These measures often have a direct impact on tourist infrastructure, e.g. dikes are increased in height and width and deplete parts of the beaches or dunes are strengthened and inhibit unrestricted view on the sea. Therefore, the local population is not always positive towards these measures and authorities have problems to communicate their necessity. The lack of awareness, that improved storm surge protection is required, is a special feature at the Baltic Sea. The Baltic Sea has a tidal range below 0.2 metres. In the Oder estuary, storm surges above 1.5 m above the mean sea level are rare and it is over a century ago that a level above 2 m was observed. Different to the North Sea, storm surges in the Baltic hardly ever cause serious damage and heavy storm surges (like in 1913 or 1874) are not remembered any more. This clearly limits the acceptance of coastal protection measures in the Baltic. Another problem is possibly the insufficient awareness rising, information and communication policy by authorities.

Why is sea-level rise often regarded as the only consequence of Climate Change? The wide range of potential Climate Change effects on our coast and our coastal ecosystems as well as their social and economic consequences are not perceived. On reason is certainly the lack of concrete knowledge. Concrete regional long-term forecasts as well as reliable regional scientific results are increasing, but still not abundant. Further, the transfer of knowledge between scientists, regional practitioners and locals is not efficient.

What we learn is that Climate Change requires an improved information, communication and education strategy, involving different media and dissemination paths. To measure the efficiency of information offers and campaign remains a major task for the future.

### 5.2 Information through the Internet: A story of success?

Internet-based tools like e-newsletters, databases, online-information modules and regional information systems are flexible, relative cheap to establish and to maintain, can easily be updated, can store large amounts of data and information permanently, are easy to access and to use, are available for regional users as well as for externals and face an increasing acceptance. Multi-language tools are certainly valuable especially in a cross-border region and improved knowledge about both sides of the border supports a joint regional identification and a corporate feeling. However, we are facing a flood of information. Due to the increasing amount of information and the easy access it becomes harder and harder to transfer information, to convince people to access and utilize information tools and really deliver messages to a target group. Internet information tools are not specific and impersonal and they usually address people with a higher educational background. They are still addressing a limited target group.

We learned that internet-based information tools have to be linked to get a critical mass. Only the critical mass, a demand driven offer and a linking with supplementary internet portals and offers ensures a permanent acceptance and utilization. Further, to be valuable for the regional population, people have to be aware of the information offers. Advertisement is required. Internet tools should not be regarded as separate, isolated products. If information shall be made available and transferred, a comprehensive information strategy and policy is needed. This includes the involvement of traditional media, own awareness rising activities in regional boards and groups as well as direct personal communication. The project ICZM-Oder maintains a regional office to promote coastal management and a sustainable development, but it is not based on a comprehensive information, dissemination and communication strategy. In this respect, improvements of the existing regional tools in the Oder estuary are still possible and necessary.

What does it mean for Climate Change? Even with respect to general topics, like a sustainable development of the coastal region it is not easy to develop, implement and maintain an information and communication strategy. Especially with respect to internet-based information tools, a critical mass of information is needed to ensure that offers are valuable and accepted. For the topic Climate Change and its consequences it means that it has to be linked to suitable existing offers, instead of establishing something new.

The major work ahead, is not to develop new internet offers or print products. What has to be improved is the efficiency and out-reach of the existing tools. Dissemination strategies and networks need to be improved for this purpose. Climate Change and its consequences for the coast should become part of the national strategy on ICZM. And the ICZM-strategy could serve as an umbrella for improved information about coastal issues and as a carrier of coastal Climate Change problems. It could possibly even help to integrate many available information and education offers.

Do we still need personal communication? All path-ways to disseminate information, which were mentioned so far, are indirect and impersonal. Are these tools and media sufficient? Does the existing regional knowledge really influence policy and decision making? We have to answer both questions with no.

Only personal communication can ensure that the information, which is needed for a decision is really meeting the demand and can be immediately used. Personal contacts and communication are still one of the most important pathways to transfer a message. Efficient and direct personal discussions and the involvement in relevant working groups are necessary. Possibly, even well focussed regional meetings, forums and workshops dealing with Climate Change could be beneficial.

## References

*Behrendt, H. & Dannowski, R.* (eds.) (2005), „Nutrients and heavy metals in the Odra river system." Weißensee Verlag, Berlin 345p.

*Hmielorz, A. & Löser, N.* (2006), „Klimawandel und seine regionalen Auswirkungen: Thematisierung and Bewusstseinsbildung in Printmedien der Odermündungsregion." IKZM-Oder Berichte 31, 36p.

IPCC (2001), „Climate Change 2001: Impacts, Adaption, and Vulnerability" – Contribution of the Working Group II to the Third Assessment Report of the Intergovernmental Panel on Climate Change; Cambridge und New York.

*Löser, N. & Sekścińska, A.* (2005), „Integriertes Küste-Flusseinzugsgebiets-Management an der Oder/Odra: Hintergrundbericht." IKZM-Oder Berichte 14, (www.ikzm-oder.de/ergebnisse_ikzm-oder_berichte.php).

*Peters, H.P. & Heinrichs, H.* (2005), „Öffentliche Kommunikation über Klimawandel und Sturmflutrisiken – Bedeutungskonstruktion durch Experten" Journalisten und Bürger, Jülich.

*Schernewski, G. & Löser, N.* (eds.) (2004), Managing the Baltic Sea, Coastline Reports 2, ISSN 0928–2734, 280 p, (www.eucc-d.de/coastline_reports.php).

*Schernewski, G., Neumann, T., Podsetchine, V. & Siegel, H.* (2001), „Spatial impact of the river Oder plume on water quality and seaside summer tourism at the south-western Baltic coast." International Journal of Hygiene and Environmental Health 204, 143–155.

*Schernewski, G. & Schiewer, U.* (eds.) (2002), „Baltic Coastal Ecosystems." Springer Verlag, Berlin.

*Wielgat, M. & Schernewski, G.* (2002), „Impact of the Odra River nutrient load reductions on the trophic state of the Szczecin Lagoon: A modelling approach." Proceedings of the International Conference 'Sustainable Management of Transboundary Waters in Europe', UNECE, 21–24 April 2002, Miedzyzdroje, Poland, ISBN: 9036954800, 347–350.

## Acknowledgement

The work was carried out within and supported by the projects IKZM-Oder (Federal Ministry for Education and Research; 03F0403A), ASTRA (Interreg IIIb; Baltic Region) and SPICOSA (European Commission, Integrated Project).

Chapter 10

# KLARA-Net – An Interdisciplinary Action Oriented Approach on Climate Change Adaptation

**Birte Frommer[1] and Lena Herlitzius[1]**

## Abstract

The man-made climate change is not a matter of scientific discussion anymore. The recent release of the 4$^{th}$ IPCC report raised the public awareness on the topic of climate change significantly. This is a great chance to switch from academic discussion to concrete political actions regarding adaptation and mitigation policies. But as no single measure will suffice to handle the complex problem interdisciplinary measures and cross-cutting actions are needed (IPCC 2001, IPCC 2007).

Various studies and approaches on possible adaptation and mitigation policies and actions have already been carried out. However, most of these studies have not fully considered the needs and perceptions of stakeholders affected by or effecting climate change. And there is no common understanding of the advantages and disadvantages of the different mitigation and adaptation options.

This paper describes the approach of KLARA-Net, a German research project, that tries to challenge the mentioned problems by building up a regional interdisciplinary network of different players all searching for adaptation strategies regarding their specific field of action.

## 1. KLARA-Net – What is it about?

Efforts to reduce the emission of greenhouse gas are currently undertaken on global (e.g. Kyoto protocol), EU, national (e.g. National Sustainability Strategy Germany) as well as on regional and local level (e.g. Local Agenda 21 processes). However, differently affected stakeholders (e.g. Industry, City and Regional Planning, Tourism and Health sector) are not equally aware of the need to define adaptation strategies regarding the already ongoing climate change (UBA 2005a, UBA 2005b).

The impacts of climate change will meet/affect the regions in Europe differently. Especially the highly urbanised and densely populated regions are facing a high vulnerability and an immense loss potential. Therefore it is important to assess the

---

1 Technische Universität Darmstadt, Fachgebiet Umwelt- und Raumplanung.

potential impacts carefully before drafting strategies und concrete adaptation techniques for different regions and different economic sectors (Jonas et al. 2005).

KLARA-Net („Netzwerk zu *KL*ima*A*daption in der *R*egion St*A*rkenburg" = Network on climate change adaptation in the Region of Starkenburg) is a pro-active bottom-up approach on regional adaptation strategies. The network takes into account (regarding) the negative effects of the climate change on the one hand but also possible positive effects or chances on the other hand.

The research area of KLARA-Net is the Region of Starkenburg in the south of the German State of Hesse. The Region of Starkenburg is part of the metropolitan area Frankfurt/Rhine Main. Starkenburg itself is a quite heterogeneous region regarding the economy (from agriculture to high-tech) and the landscape (from the low mountain range of the Odenwald to the urbanised areas of the Rhine rift valley) of its districts. Hence, different consequences of climate change have to be expected and a regional adaptation strategy has to consider diverse spatial circumstances. The heterogenic consequences of climate change also require answers and actions from various stakeholders. Besides, the borders of the network are rather weak to spread the approach to additional stakeholders outside the Region of Starkenburg.

Thus KLARA-Net is focusing on the establishment of a regional stakeholder network that works out fields of action and particular measures for climate change adaptation. Additionally, KLARA-Net is undertaking actions to raise public awareness on climate change and its regional impacts.

KLARA-Net uses the information and data collected within the work on the „Integriertes Klimaschutzprogramm 2012" (INKLIM 2012) = Integrated climate protection programme 2012 of the state of Hesse (HLUG 2005b). Additionally, KLARA-Net is using data from climate models offered by the Max Planck Institute for Meteorology. Especially the detailed research on the consequences of climate change for the state of Hesse is needed to further develop adaptation measures for the region.

KLARA-Net is funded by the Federal Ministry of Education and Research (BMBF) with a project time of one year (6/2006–5/2007). The research project KLARA-Net is following on the one hand the aims of the Framework Programme „Research for Sustainability" for a sustainable, innovative society („fona") and on the other hand the programme „Research for Climate Protection and Protection from Climate Impacts" („klimazwei") (BMBF 2004a, BMBF 2004b). A continuation phase is planned to implement measures that are identified during the first project phase. The section „Environmental and Spatial Planning" at the Technische Universität Darmstadt is moderating and coordinating the process and evaluates the network scientifically.

Within this paper we are going to describe how KLARA-Net was established. Also, the process of networking will be analysed. First results on concrete regional measures to deal with impacts of climate change will be presented as well as an analysis of the cooperation and communication structure of the network.

## 2. Networking – The main element of KLARA-Net

The ongoing climate change affects a wide range of branches in different ways: Farmers may need to adapt their crop and current cultivation methods to rising temperatures and changed precipitation. The tourist sector has to face a changing attractiveness of the regions when e.g. former winter sport resorts do not longer have reliable snow conditions but a better climate for summer activities. The building industry might develop new markets when providing improved insulation or climatisation techniques. These examples give only a small insight into the fields of action that need to be taken into account (Zebisch et al. 2005).

But adaptation strategies and techniques to tackle those challenges are either not yet developed – or if they are developed in theory, they are not yet implemented in practice. This is mainly due to the fact that players responsible to realise such adaptation strategies are not yet aware of the risks and chances that climate change may bring to them. The specific regional impacts and the different scenarios of climate change are seldom known, and economic consequences are ambiguous (Elsner et al. 2005).

The KLARA-Net project therefore has three main goals:

1. To raise the awareness for the general risks and chances of climate change
2. To discuss the possible impacts of climate change regarding the Region of Starkenburg
3. To define concrete adaptation strategies and techniques concerning this region.

By using an interdisciplinary and action-oriented approach of regional networking KLARA-Net opens the possibility to confront regional stakeholders with the impacts (negative as well as positive) of climate change and (by doing so) to develop fields of actions needed in the region. The multidisciplinarity of the network shall also improve the understanding of regional interdependencies that require cooperation. The topic „networking" is also regarded as highly relevant by the Federal Ministry of Education and Research, the funding institution of KLARA-Net.

Within the KLARA-Network three different levels of stakeholder involvement can be distinguished. First there are the „active players" within the network, secondly there are players outside the network working as knowledge or vision multipliers, and thirdly there is the general public.

Fig. 1: Levels of involvement within the network of KLARA-Net

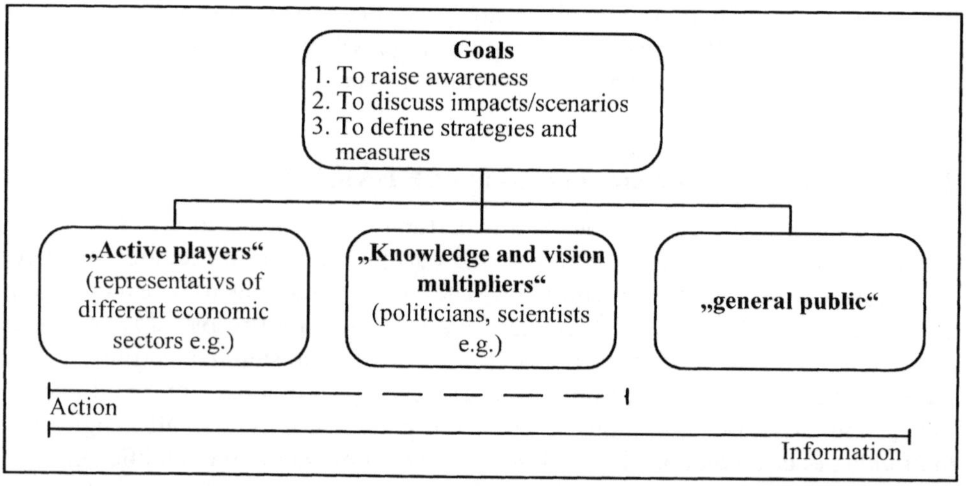


The „active players" within the network are members of different economic sectors that have been recruited to take part in four different thematic working groups. They are the active part of the network mainly dealing with the possible regional impacts of climate change (goal no. 2), and they determine adaptation strategies and techniques concerning their own fields of action (goal no. 3). A detailed description of the constellation and motivation of these players is part of chapter 3.

Additionally there are players like scientists from related research projects, politicians or other regional important decision-makers who are not taking part in the working groups but who are „knowledge or vision multipliers" asked to spread the idea of KLARA-Net within further networks or organisations. They are contacted and informed via conferences and symposiums or – regarding the politicians – via informal meetings.

The overall goal no. 1 to raise the awareness on climate change is covered by all three levels of involvement and related target groups. The „active players" and the additional knowledge and vision multipliers will disseminate the experience gained from KLARA-Net to their organisations.

## 3. Constellation and motivation of the players involved

The establishment of a regional stakeholder network means a quite new attempt to strengthen the public awareness and to start the implementation of action needed regarding climate change adaptation. To have a successful process the constellation and motivation of the players acting within this network is crucial. Although drafted as open process KLARA-Net started with a preselection of relevant players. The selection criteria, the constellation and the motivation of the players are subject of the following depictions.

## 3.1 Drafting the project structure

Climate change and its impacts are very complex matter. When drafting the project structure this complexity had to be handled carefully. Many economic sectors needed to be taken into account, and their vulnerability had to be analysed. Therefore, the project agreed on a structure that is firstly focussing on sectors that are characterising the region. Secondly, sectors were chosen that are highly vulnerable and have the potential to engage a wide range of stakeholders. The possibility of chances due to climate change was also a criterion. Furthermore the private sector was sought to be included in the network. As „awareness raising" is a central goal of KLARA-Net the chosen subjects should also be of special interest for the general public and the media.

To sum up the selection criteria for actors/working groups of the network have been:

- Coverage of spatial heterogeneity of the region
- Highly vulnerable sectors
- Integration of sectors that may profit from climate change
- Participation of the private sector
- Interest of the public

The decision to involve the health sector e.g. was taken because of its close relation to the public (awareness). Especially the involvement of actors from tourism allows discussing chances of climate change in the region. Chances may not be directly visible but partly hidden under measures of adaptation, e.g. new technologies, new marketing aspects and regional cooperation aspects (see figure 2).

Besides, KLARA-Net has to be understood as an open process that is also possible to be enlarged when important issues seem to be missing. To identify and integrate such issues, the process of KLARA-Net is undergoing a continuous evaluation. The evaluation also reviews the structure of the network and the satisfaction of all involved players (more on the project evaluation in chapter 4.4).

## 3.2 Step by step – Establishment of the regional network

The knowledge about important and engaged regional players can be stated as one decisive issue while starting, coordinating and moderating such a process. However, such knowledge is often based on former networking processes. In this case, negative experiences of cooperation might confront actors once again. Also, regional barriers might have resulted from former networks or cooperation that will also be transferred into the new cooperation process. Thus, the knowledge of regional cooperation in past and present times when building up new networks is both helpful but also bears certain risks.

- Knowledge about the existence of regional networks in past and present times helps to understand the specific character of regional cooperation and to define key persons.
- Knowledge about possible negative experiences between regional players should not increase barriers but should rather help to find new ways to cooperate.

Against this background regional stakeholders from different sectors and political levels were informed about the goals of KLARA-Net and asked to participate already during the application for financial support of KLARA-Net. As expected, many stakeholders confirmed their interest on participating within the network, partly as active members within specific working groups or more in the background as multiplier or in a more political role. With that a first group of interested people had been found.

The next step to constitute the proposed working groups was to talk to interested stakeholders in bilateral meetings to discuss their expectations from and possible input into the network. Within this phase of the project, first fields of action and specific knowledge of stakeholders could be collected. One surprising fact was that many stakeholders already thought about fields of action and possible measures for adaptation but they did not define them clearly as adaptation measures.

The kick-off event of KLARA-Net in July 2006 was the official start of the project. A wide range of regional stakeholders was invited. Those persons who already confirmed their interest before were asked to shortly present their specific interest to join the network. The individual motivation could be directly discussed within the plenum. Additionally, visitors of the kick-off event were asked to fill in a questionnaire that aimed to get an overview about interests to join the network and current knowledge about climate change. The results of the questionnaires are embedded within the process evaluation of KLARA-Net (see chapter 4.4).

To establish a structure for the network, the actors were thematically clustered into four groups. The actors were selected based on the criteria mentioned above (see chapter 3.1), i.e. they cover different sectors of the region with varying vulnerability to climate change. The four working groups were suggested to and discussed with the stakeholders in the kick-off event and then agreed upon (Frommer and Herlitzius 2006). The presently participating players fit rather well to the proposed constellation in the design phase of the project (see figure 2). The figure only shows an overview about the main groups of stakeholders.

Finally, the following four working groups started their work in October 2006.

- Agriculture, forestry and viniculture
- Building economy, water economy and planning
- Health
- Tourism

Fig. 2: The interdisciplinary working groups of KLARA-Net

| Working Groups | **Agriculture, Forestry and Viniculture** | **Building Industry, Water Management and Planning** | **Health** | **Tourism** |
|---|---|---|---|---|
| Goals | Adjustment of product range<br>Adjustment of production methods | Adaptation of regional and local planning<br>Advancement of materials and building designs | Precaution measures<br>emergency plans<br>early warning systems | Adaptation strategies relating to different tourists attractions |
| Stakeholders | Farmer's Union<br>Forestry offices<br>NGOs<br>Forestry associations<br>Winegrowers<br>Consumer counselling<br>City administrations<br>Research institutions<br>Consultants | Chamber of craft<br>Chamber of commerce and industry<br>Producer of building materials<br>Insurances<br>Consumer counselling<br>City administrations<br>Research institutions<br>Planning authorities (lovcal/regional) | Hospitals and nursing homes<br>Health insurances<br>Medical practioneers<br>Local health authority<br>German meteorological service<br>City administrations<br>Research institutions<br>Consultants | Associations of tourism<br>Association of hotel business<br>Chamber of commerce and industry<br>Regional marketing<br>City administrations<br>Research institutions<br>Consultants |

The goal of these working groups is to define fields of action and adaptation measures that are of high relevance for the region. Many of those actions and measures concern not only one but several working groups or sectors of the region. To identify whether measures are cross-cutting or rather mono-sectoral, actions and measures are presented to the other working groups, and a steering committee oversees the whole process to ensure good coordination.

Fig. 3: Organigram of KLARA-Net

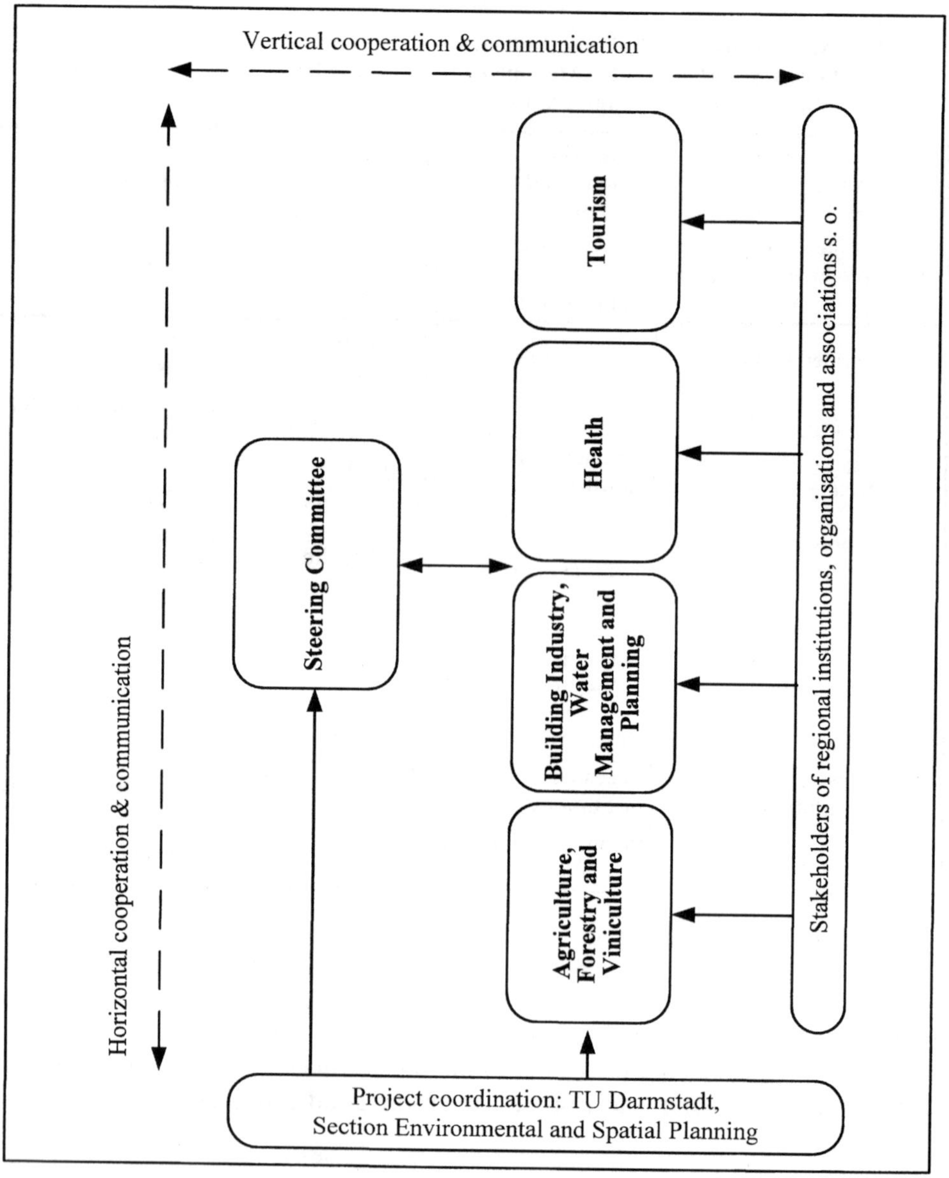


## 4. First results of KLARA-Net

The network KLARA-Net can now present first results from its active work since October 2006. Additionally, the network can present several lessons learnt on spreading the approach and raising public awareness. The overall impacts of such a network are difficult to measure quantitatively but it is possible to describe impacts and tendencies in a qualitative way.

### 4.1 The four working groups of „active players"

The first meetings were characterised by an ample collection of fields of action and partly already quite specific measures which actors judged as important to integrate into a strategy of adaptation in the region. On average, 10 persons took part in every working group within the first round. The collection of ideas showed that knowledge about climate change and its impacts varied widely among group members. The level of knowledge needs to be balanced to discuss about measures of adaptation. In addition, it can be stated that the level of awareness is strongly connected with the individual vulnerability of different sectors. For example, representatives of the highly vulnerable agricultural sector had already concrete ideas about adaptation measures (like changing parts of the crop e.g. growing cantaloupes and kiwi fruits), whereas representatives of the tourism sector first needed to be informed about possible regional impacts of climate change. Due to the recent hot summers the „Health" working group was aware of severe climatisation problems meeting hospitals and residential homes for elderly people e.g. But new to some of them was the increasing risk of tropical or vector based diseases in the region (Rahmstorf & Schellnhuber 2007). And although the water and infrastructure sector has perhaps the deepest awareness to climate variability many possible impacts of climate change were new to many members of the working group „Building economy, water economy and planning" (see figure 3).

By following an interdisciplinary approach within KLARA-Net through exchange between the working groups, exchange and dissemination of information is fostered. The fact that members know each other partly from other projects contributes to efficient discussions. The mostly discussed action fields are public awareness raising, adaptation of technologies and adapted methods concerning the use of natural resources, measures on flood prevention and conservation of water quality. Besides, the inclusion of possible chances or profits that go along with climate change helped to view the consequences of climate change not only in a negative way. This also increased motivation for the conceptual work.

One interesting aspect is that collected fields of action are mostly known in the context of approaches of (regional) sustainable development. Thus, measures to adapt to climate change are strongly connected to already known aspects of sustainability. This finding can be interpreted as follows:

- Existent approaches on (regional) sustainable development need to be taken up also within the context of climate change (protection and adaptation).
- Additional fields of action need to be clearly defined and put into context with sustainable development.

A common subject of the working groups was the resource water. In addition to measures regarding flood management and irrigation, actions were suggested that include subjects that need to be dealt within cross-cutting action. The identification of cross-cutting issues like the future role of water management opens fields of action that involves players between the working groups. The resource water concerns not only the further distribution of drinking water or irrigation in the region, but also lakes as tourist attractions and hygienic aspects come up in this context. Thus, cross-cutting issues were found and identified as strategic for the region in that sense that they lead to a more coordinated adaptation in the region (see chapter 4.2).

- Cross-cutting issues open a new understanding of regional fields of action and therefore support regional common action. They are identified as a chance for the region to strengthen regional coordination.

Thus, the work on adaptation on climate change started regional fields of action that combine already identified actions for regional sustainable development with new and especially cross-cutting actions. During several working groups, a discussion came up on how to use current regional funding programmes for climate change adaptation issues. The adjustment of such programmes needs to be addressed by the network in order to connect already existing approaches with new aspects. In this context the role of regional planning authorities might be an important one to anchor aspects of climate change on a regional level. Both, the local and regional level needs to be regarded and responsibilities have to be shared. Within the current working groups the importance of local action is discussed to firstly, strengthen the awareness of needed adaptation measures and secondly to use the implementation level to start with pilot actions in the region that might motivate additional stakeholders. The regional level is mostly mentioned to find a way to adapt current and future regional planning actions (plans and programmes) to a changed climatic situation in future. Especially, long during planning processes are discussed in that context and therefore a fast definition of adaptation measures should be aimed.

### 4.2 Steering Committee of „active players"

During the project so far, the complexity of climate change adaptation became clear to all members. The network also already helped to understand the needs of different sectors. In addition, the cross-cutting work identified interdependencies between the different sectors that clearly showed the need for common action in the region. Especially, the heterogenic character of the landscape and therefore the heterogenic patterns of land use in the Region of Starkenburg are interesting concerning climate change aspects. Regional interdependencies open new aspects of needed regional cooperation. Parts of the region have to deal with different consequences of climate change (increasing risk of floods in the lowlands of the Rhine rift valley, worsening condition for the winter sport industry in the mountainous part of the Odenwald, improving climatic conditions for the viniculture at the Bergstraße) and therefore the regions needs to get a general view about its potential chances and losses in future.

The role of the Steering Committee is to identify the cross-cutting issues which are connected to several fields of action or measures proposed by the four working groups. Members of the Steering Committee are representatives of the working groups. The constitution of this committee opens the possibility of knowledge exchange and strategic considerations. The Steering Committee analyses the fields of action defined by the working groups and clusters them into strategic approaches for the region. Those strategic approaches are then presented to the working groups for comments and agreement (see figure 3). To give an example, the decision to concentrate at the first stage of the project on the future management of the resource water (adaptation of irrigation measures, water storage etc.) was proposed by the Steering Committee intending to connect measures within the four working groups (see chapter 4.1). With that, common fields of action can be defined and their implementation shall be pushed forward. The Steering Committee is following the strategy to collect and structure proposed measures but also to identify measures that could be strategic in that context that they are functioning as pilot projects to start with implementation in the region and with that to strengthen the public awareness.

Within the Steering Committee also aspects are discussed that focus on different development scenarios of our society. Today, the demographic change and its consequences are a key issue in Europe. The idea is to observe parallel developments that may concern or influence proposed actions of KLARA-Net.

### 4.3 Public awareness

Public awareness is a central issue to anchor the importance of climate change in the region.

Up to now, actions concerning climate protection are well known in the region and several active players are involved in organisations or processes dealing with climate change and reduction of greenhouse gases (e.g. regional Agenda 21 processes). However, adaptation of climate change opens new aspects and a kind of rethinking. Therefore, KLARA-Net tries to transmit the importance of adaptation but also remind people of the importance of climate protection.

The network KLARA-Net contributes to awareness raising on climate change and adaptation measures regarding both its active participants and the general public. The ongoing awareness raising among the public is of course not easy to assess, especially considering the short duration of the project.

Related to awareness raising among participants, regular evaluation steps of the project show that the majority of them could broaden their knowledge on climate change and action needed in the region. Additionally, they are more aware of the different sectors affected by climate change due to the continuous exchange between the four working groups of KLARA-Net. The network as such is flexible, which allows for its further extension when already active participants engage further actors to join the network („snowball effect").

Beside the rising awareness of network participants, the general public in the region is also informed about climate change and its impacts through KLARA-Net. To reach the „general public" various PR measures are carried out using different media: For example, KLARA-Net is presented at regional events like Agenda 21-forums or regional fairs. The high interest of the media in KLARA-Net (since starting the project there have been various requests to present KLARA-Net in TV, radio and the local and regional press) shows an already increased awareness. During the warm winter 2006/2007, journalists informed about climate change and its consequences not only on a global, but also at a regional or local scale. Several interviews about the objectives and first results of KLARA-Net informed the public about regional consequences and aspects of concern. The chosen scale of the project gives people a chance to understand what climate change means for the region. And apart from that the broad dissemination of the project idea and first results awoke the interest of politicians that now closely follow the development of KLARA-Net and its proposals. Local and regional parliamentary parties asked for project presentations to get an overview about action needed in the region. Besides, politicians also participated in KLARA-Net events like the regional forum held in autumn 2006 or a scientific symposium to be held in spring 2007.

### 4.4 Process evaluation of KLARA-Net

The process of KLARA-Net is monitored by a self-evaluation that is inviting regional players to mention their opinion about the project. Up to now two questionnaires were analysed: one was distributed at the kick-off event in July 2006, the other at the public regional forum in November 2006[2]. A third questionnaire is in preparation which will also include proposals for a concept for a second phase of KLARA-Net. The questionnaires ask for judgements of active participants of the network, stakeholders who merely follow the events as well as multipliers who consider the process from a broader point of view. In addition to that, the regular contact to working group members provides a permanent exchange of suggestions.

---

2 The documentations of both KLARA-Net events are published on the project homepage: www.klara-net.de.

The main results of the questionnaires can be summarised as follows:

- Although most of the participants felt already well informed about climate change, they could broaden their knowledge during the process of KLARA-Net so far.
- Several institutions are already working on matters of climate change including measures of public awareness raising (e.g. exhibitions, seminars), but they are mostly concentrating on climate protection. Adaptation issues are mostly lacking.
- The constellation of the four current working groups is considered to be appropriate
- An increasing active involvement of politicians is required to put proposals and measures into action (action-oriented approach)
- The regional scale of KLARA-Net seems to be appropriate, with the number of network members and topics being manageable.

In addition to questions addressing the process of KLARA-Net, the second questionnaire also addressed the three different roles of man in the context of climate change. The stakeholders were asked to give their personal estimations if they feel more like:

- a person causing climate change
- a person being affected by climate change
- an active player coping with the impacts of climate change.

Most of the interviewees understand themselves as highly affected, but less causing ones. With regard to the idea of KLARA-Net to identify adaptation measures, it is interesting that many persons see little possibilities to cope with the problems and impacts of climate change. There is obviously a high demand to define fields of action on adaptation to climate change.

To sum up, the current process of KLARA-Net fits quite well to the demanded fields of action of involved players. Mentioned fields of action were strongly connected to the future management of natural resources, especially the resources water (future water management, keeping the quality of drinking water etc. and soil (aspects of sustainable land use, infiltration etc.). Additionally, actors are motivated to define fields of action that might imply chances for the region (e.g. new tourist attractions, new cultivations in the agricultural sector etc.). The interdisciplinary approach covers a broad range of topics that should be tackled in the context of climate change adaptation, but stronger participation of decision-makers is required by the stakeholders to come to real action in the region and to implement the proposed actions.

## 5. The importance of system-oriented thinking

Many scientific reports on the impacts of climate change have already been written. Multiple measures have been identified, too. But the concrete implementation is still missing in most of the cases. KLARA-Net therefore tries to fasten the implementation of measures by involving regional players who are voluntarily spending time. Players will accept measures developed by them selves more then measures proposed from top-down. As the players are responsible for the implementation themselves in many cases the pressure for action and the motivation should be higher. Additionally, the integration of scientific expertise into regional action seems to be one central issue for successfully adapting regions in Europe to climate change. KLARA-Net is following a „bottom-up" approach that is using scientific expertise in the region (e.g. new results of climate modelling projects by the Max-Planck Institute for Meteorology and results of INKLIM 2012, the „Integrated climate protection programme 2012" of the state of Hesse) to develop action-oriented adaptation measures (HLUG 2005b).

As already shortly described above, the interdisciplinary approach of KLARA-Net opens an interesting view to fields of action concerning the adaptation of climate change. Especially the identified interdependencies of different branches motivate the actors to find solutions that satisfy more than just one interest. To strengthen that, KLARA-Net is focussing on adaptation measures that pick up those fields of action that reflect the cross-cutting issues of branches in the region. The approach has the potential to strengthen the common interest of actors and with that to achieve joint action in the region.

One example is a planned project of a specific hiking trail that is intended to inform hikers about natural springs in the region and their possible loss in future because of the predicted climate change (public awareness raising). Besides, aspects on health will be included within this hiking trail (regular water consumption, breaks about noon etc.). The idea was born within the working group „tourism" but raised also discussion in other working groups (health, water management etc.). The approach on thinking in interdependencies (system-oriented) allows a much more intensive exchange of knowledge between the different working groups and with that the overall task on climate change adaptation is being based on common interest.

Furthermore, the aspect of time is quite important when working on climate issues. The global climate change is a rather slow process that has neither start nor an ending point. This fact makes it hard to define the right time to begin with the implementation of adaptation measures. Does it exist, the ideal time to start with climate change adaptation? The aspect of time does confront branches in different ways, for instance the forestry has to plan within long-term perspectives that aren't that flexible to change (e.g. domestication of forest species). The understanding of different time perspectives of branches needs to be handled carefully to be able to define adequate time for action.

## 6. Spreading the regional network approach

Dealing with climate change adaptation on an action-oriented and on regional scale is quite innovative. The chosen bottom-up network approach gives one possibility to prepare the region on climate change. Measures are more concrete on a regional then a national level and the inclusion of politicians and other decision-makers leads to better commitment.

Therefore it is intended to discuss and spread the approach also outside the network. Thus, one part of KLARA-Net is to disseminate the approach and to exchange knowledge within the region but also with other regions of Germany and Europe.

Within the region one part of the involved stakeholders is following the role as multipliers (cp. chapter 2). Those persons connect KLARA-Net with parallel existing networks (e.g. projects on climate protection) and with resource persons to integrate their knowledge and experiences. Additionally, they take part in events of KLARA-Net that aim to inform regional stakeholders and the general public about climate change. But also active players are spreading the approach and anchor the issue of climate change adaptation within their institutions.

Today, several conferences and seminars offer platforms for the exchange of adaptation strategies such as KLARA-Net. Hence, KLARA-Net was presented e.g. during the 2. ASTRA Conference (ASTRA= Developing Policies and Adaptation Strategies to Climate Change in the Baltic Sea Region") in Gdansk (Poland) in autumn 2006, where adaptation strategies represented one focus of discussion.

## 7. Conclusion

The ongoing climate change asks for coordinated action – mitigation as well as adaptation – in regions all over the world. KLARA-Net uses the approach of interdisciplinary action to define a common action within the Region of Starkenburg. The multi-faceted global as well as the regional consequences of climate change can be collected and discussed between players within the regional network. The focus on defining cross-cutting actions strengthens the relevance of common action in the region. The regional interdependencies become visible and form in many cases interesting starting points of regional action.

Several institutions are supporting the idea of KLARA-Net in the region. Institutions of the state of Hesse encourage the network to go into a second phase. The character of support is partly content-oriented (participation in events of KLARA-Net etc.) or financial. Within the planned second phase of KLARA-Net one decisive difference is intended to be realised: The direct involvement of decision-makers will be added to the network to fasten the implementation of proposed actions. The further anchorage of matters on climate change in the region needs support by regional players but also by decision-makers who encourage stakeholders for implementation of needed action. The Federal Ministry of Education and Research counts on the interdisciplinary character of the current network and showed interest to further support common regional action and public awareness raising in the region.

The known economic consequences of climate change increase the urgent need for action. This pressure needs to be used to speed up implementation of measures for adaptation to climate change. Implementation of proposed actions will adapt the region to changed climatic situations now and in the future. The pressure for action will also contribute to finally realising measures for sustainable development that were developed long ago. Especially regional planning authorities are required to increasingly include aspects of climate change within their plans and programmes.

## References

BMBF, Bundesministerium für Bildung und Forschung (2004a), Forschung für die Nachhaltigkeit. Rahmenprogramm des BMBF für eine zukunftsfähige innovative Gesellschaft. Bonn/Berlin.

– (2004b), Forschung für den Klimaschutz und Schutz vor Klimawirkungen. Ein Beitrag zum BMBF-Rahmenprogramm „Forschung für die Nachhaltigkeit". Bonn/Berlin.

*Elsner, W.* et al. (2005), Klimawandel und regionale Wirtschaft, Vermögensschäden und Einkommensverluste durch extreme Klimaereignisse sowie Kosten-Nutzen-Analysen von Schutzmaßnahmen, Am Beispiel der nordwestdeutschen Küstenregion, Peter Lang GmbH Europäischer Verlag der Wissenschaften, Frankfurt am Main, pp. 9–16

*Frommer, B. & Herlitzius, L.* (2006), Klimawandel als Chance?! Dokumentation der Auftaktveranstaltung des Projekts KLARA-Net am 10.07.2006. Available www.iwar.bauing.tu-darmstadt.de/umwr/deutsch/forschung/KlaraNet/KLARA-Net_Dokumentation_Auftakt.pdf

HLUG, Hessisches Landesamt für Umwelt und Geologie (2005b), Integriertes Klimaschutzprogramm Hessen, INKLIM 2012, Projektbaustein I1: Klimawandel und Klimafolgen in Hessen, Abschlussbericht, Wiesbaden

IPCC, Intergovernmental Panel on Climate Change (2001), Third Assessment Report.

Climate Change 2001, Cambridge University Press, Cambridge, UK.

– (2007), Forth Assessment Report.

Climate Change 2007: The physical science basis, Summary for Policy Makers. Available www.ipcc.ch/SPM2feb07.pdf

*Jonas, M., Staeger, T. & Schönwiese, C.D.* (2005), Berechnung der Wahrscheinlichkeiten für das Eintreten von Extremereignissen durch Klimaänderungen – Schwerpunkt Deutschland. Kurzbericht zum Forschungsvorhaben 201 41 254 im Rahmen des Umweltforschungsplans des Bundesministers für Umwelt, Naturschutz und Reaktorsicherheit.

*Rahmstorf, S. & Schnellhuber, H.J.* (2007), Der Klimawandel, Diagnose, Prognose, Therapie, 4.Auflage, C.H. Beck oHG, München, pp. 79–80

UBA, Umweltbundesamt (2005a), Berechnung der Wahrscheinlichkeiten für das Eintreten von Extremereignissen durch Klimaänderungen – Schwerpunkt Deutschland, in: Climate Change 07/05. Dessau.

– (ed.) (2005b), Klimawandel in Deutschland. Vulnerabilität und Anpassungsstrategien klimasensibler Systeme, in: Climate Change 08/05. Dessau.

– (2005c), Hintergrundpapier: Klimafolgen und Anpassung an den Klimawandel in Deutschland – Kenntnisstand und Handlungsnotwendigkeiten. Dessau September 2005.

Wuppertal Institut für Klima, Umwelt, Energie (2004), Hintergrundpapier: Anpassung an nicht mehr vermeidbaren Klimawandel. Wuppertal 3.12.2004.

*Zebisch, M., Grothmann, T., Schröter, D., Hasse, C., Frisch, U. & Cramer, W.* (2005), Klimawandel in Deutschland. Vulnerabilität und Anpassungsstrategien klimasensibler Systeme. Kurzbericht zum Forschungsvorhaben 201 41 253 im Rahmen des Umweltforschungsplans des Bundesministers für Umwelt, Naturschutz und Reaktorsicherheit.

**Further reading**

*Ackerman, F. & Stanton, B.* (2006), Climate Change – The Cost of Inaction. Friends of the Earth. Available www.foe.co.uk/resource/reports/econ_costs_cc.pdf.

*Berkhout, F.* (2003), How Can Business Adapt to Climate Change? Conference Summary. Available www.tyndall.ac.uk/events/past_events/adapt.doc.

*Berkhout, F.* et al. (2004), Business and Climate Change: Measuring and Enhancing Adaptive Capacity. Tyndall Centre. Available www.tyndall.ac.uk/research/theme3/final_reports/it1_23.pdf.

BMBF, Bundesministerium für Bildung und Forschung) (n.y.): Forschung für den Klimaschutz – Stand und Perspektiven, in: BMBF Publik. Bonn/Berlin.

– (2005), Nationales Klimaschutzprogramm 2005. Berlin.

*Easterling, W.E.* et al. (2004), Coping With Global Climate Change. The Role of Adaptation in the United States. Pew Center. Available www.pewclimate.org/docUploads/Adaptation.pdf.

*Firth, J. & Colley, M.* (2006), The Adaptation Tipping Point – Are UK Businesses Climate Proof?. Acclimatise and UKCIP, Oxford. Available www.cdproject.net/download.asp?file=CDP4_FTSE350_Adaptation_Report.pdf.

*Heintze, C.* (2002), Die Zukunfts-Blockade: Klimawandel, BSE, Armut, Terrorismus – warum in der Gesellschaft kollektives Vorsorgelernen misslingt, Logos Verlag, Berlin.

HL, House of Lords (2005), The economics of climate change. Available www.publications.parliament.uk/pa/ld200506/ldselect/ldeconaf/12/12i.pdf.

HLUG, Hessisches Landesamt für Umwelt und (2005a), Jahresbericht 2004 des Hessischen Landesamtes für Umwelt und Geologie, Wiesbaden.

*Kress, A.* (2006), Maßnahmen zur Anpassung und Vermeidung – ein integrierter Ansatz. Available www.amica-climate.net/fileadmin/amica/inhalte/dokumente/AMICA_UBA_ Vortrag_ final_dt.pdf.

LCCP, London Climate Change Partnership (2006). Adapting to Climate Change: Lessons for London. Greater London Authority, London. Available www.london.gov.uk/climatechangepartnership/docs/adapting-climate-change-london.pdf.

*Rack, E.* (1999), Klimawandel & Klimaschutz als Thema kommunaler Öffentlichkeitsarbeit; die Darstellung von Umweltwissen und Umweltschutzmaßnahmen zur Bildung von Umweltbewusstsein; mit einer Analyse der Kampagne „Klimaschutz Heidelberg-gemeinsam gegen dicke Luft (Natur-Raum-Gesellschaft; Bd. 2), Selbstverlag Institut für Didaktik der Geographie, Frankfurt am Main.

*Scheraga, J.D.* (1998), Risk, opportunities and adaptation to climate change. In: Climate Research, 10, S.85–95.

*Smith, J.B.* (2003), Conceptual basis for adaptation technologies. Available www.resourcesaver.com/file/toolmanager/O105UF1355.pdf.

*Stern, N.* (2006), The economics of climate change: the Stern review. Cambridge University Press, Cambridge, UK.

*Stock, M.* (2003), Chancen und Risiken von Regionen im Klimawandel: Welche Strategien kann die Wissenschaft ableiten. Abgerufen am 28.09.06, Available www.pik-potsdam.de/~stock/paper/arl_ak_iia3_stock.pdf.

*Tompkins, E.L.* et al. (2005). Linking adaptation research and practice. Available www.defra.gov.uk/science/project_data/DocumentLibrary/GA01077/ GA01077_2664_FRP.pdf,

UNEP FI (2006), Adaptation and Vulnerability to Climate Change: The Role of the Financial Sector. Available www.unepfi.org/ fileadmin/ documents/ CEO_ briefing_ adaptation_vulnerability_ 2006.pdf.

*Viner*, *D.* (2006), Climate Change and the European Countryside: Impacts on Land Management and Response Strategies. Available www.cru.uea.ac.uk/link/new_link/clio/CLIO_Main.html.

*Warren*, *F.J.* (2004), Climate Change Impacts and Adaptation: A Canadian Perspective. Available www.espp.msu.edu/climatechange/canadaadaptation.pdf.

Chapter 11

# Climate Change, Featuring the ACCROTELM Project: Dissemination of a European RTD Project by Film and DVD

**F.M. Chambers, J.R.G. Daniell, Sally A. Brain**

## Abstract

The European Commission part-funds RTD (Research and Technological Development) projects, and is keen to see wide dissemination of project results. A recently conducted RTD project called ACCROTELM (**A**brupt **C**limate **C**hanges **R**ecorded **O**ver **T**he **E**uropean **L**and **M**ass) investigated two 'proxy-climate' archives: peat bogs (across Europe) and lakes (through lake-level changes in West-Central Europe). RTD projects commonly hold a final meeting to engage with a wider community, though in practice a sizeable audience is difficult to achieve. Also, production of a succinct project communiqué that avoids 'mixed' messages might suppress minority scientific views, or result merely in an anodyne overall message. An innovation in dissemination from this project was the production of a 35-minute film, entitled '*Climate Change, featuring the ACCROTELM project*'. It contains interview footage with leading palaeoclimatologists, to voice the scientific concerns, concepts and uncertainties; illustrates methods used in ACCROTELM to obtain 'proxy' records of the past climate of Europe; and emphasises the value of palaeoclimate data for testing and refining climate models. The film, which is pitched at a level accessible to the intelligent layperson, is now hosted permanently on the Website of a Dutch TV/Radio station. Currently, the film is incorporated within a DVD containing additional interview footage, but in a later version will be linked to wider educational resources concerning climate change, which are intended to be 'layered' for the teaching of climate change at undergraduate and postgraduate levels. Despite evident imperfections, initial feedback on the film and DVD is encouraging, with requests for further copies received.

## Introduction

The European Commission supports European science partly through Research and Technological Development (RTD) projects, based on Framework directives. In such projects, an increasingly important element, which in later Framework Calls constitutes part of the evaluation of project applications, is the plan for dissemination of results. The Commission has various formal mechanisms for dissemination, via its associated website (CORDIS 2007), but also encourages projects to engage directly with stakeholders and the wider community. This is laudable; but wide dissemination outside the immediate science or technological community can be difficult to arrange in practice, as outreach through end-of-project meetings usually will involve only a relatively small group of people external to the consortium.

We recently co-ordinated a Fifth Framework RTD project called ACCROTELM (**A**brupt **C**limate **C**hanges **R**ecorded **O**ver **T**he **E**uropean **L**and **M**ass), over the 3.5 years from April 2003 to end-September 2006 (ACCROTELM 2007a). The project involved the formal participation of some 30 scientists based in 12 universities from seven EU countries (tab. 1), and the voluntary participation from scientists at three other universities – two from the UK and one in Mexico. Dissemination to scientists within the discipline and in closely allied fields has been via conventional means – a range of publications in international peer-reviewed journals, conference proceedings and abstracts (ACCROTELM 2007b, c, d).

Tab. 1: Partner institutions in ACCROTELM

| | **Partner institution** | **Country** | **Lead Scientist** |
|---|---|---|---|
| 1 | University of Gloucestershire | UK | Prof. Frank Chambers |
| 2 | Universiteit van Amsterdam | The Netherlands | Dr Bas van Geel |
| 3 | Tallinn Institute | Estonia | Dr Edgar Karofeld |
| 4 | University of Helsinki | Finland | Prof. Atte Korhola |
| 5 | University of Joensuu | Finland | Dr Jukka Alm |
| 6 | University of Plymouth | UK | Prof. Dan Charman |
| 7 | Trinity College, University of Dublin | Ireland | Dr Fraser Mitchell |
| 8 | Université de Franche-Comté | France | Dr Michel Magny |
| 9 | University of Hannover | Germany | Prof. Dr. Hansjörg Küster |
| 10 | University of Bristol | UK | Prof. Richard Evershed |
| * | University of Groningen | The Netherlands | Dr J. Hans van der Plicht |

*Subcontractor to Partner 2

Engaging with the wider community proved more challenging. In the first three months of the project a public domain website was established (ACCROTELM 2007a), so giving some freedom in terms of content, including use of font, colour, logos, etc. We do not, however, have statistics on that server concerning the number of 'hits' nor their origin, so we cannot monitor the effectiveness of the Internet as a means of dissemination from our project website.

Our original schedule incorporated an 'End-User Forum', which was intended to help disseminate the results of our project to those who might benefit from our findings. The precise format for such a forum was unspecified, although we had envisaged it as a meeting in a central, accessible location in Europe (such as Amsterdam, or perhaps Brussels) to which end-users would be invited. End-users were originally identified as other scientists (including climate modellers), policy-makers and planners. However, as the project developed, it became evident that there was a much wider constituency that was now increasingly concerned about climate change and for whom our project may be of interest. In view of this, we began to re-conceptualise how we might disseminate information concerning our project, and how we could set our project aims and its findings in the context of current scientific and public concern over future climate.

At the formulation stage of ACCROTELM, individual scientists within the project team held differing views on the relative success of previous attempts (such as by Mann et al. 1998, 1999) to produce credible reconstructions of past climate changes from various archives. In any dissemination of an overall synthesis from the project it was important to acknowledge the scientific uncertainties. While some individual views may have changed over the course of the project, there nevertheless remained by the close a healthy range of opinion. It was recognised that including a diversity of scientific opinion in communication of complex project results to a lay audience might be difficult, without sending out a confused or 'mixed' message. However, to distil into a short project communiqué one overall message, would risk either that it would be biased in favour of a majority view, or be so generalised as to be anodyne. Some other vehicle was required, which could set the context for the project, permit a range of opinions to be expressed, explain the scientific techniques used to address the topic, and still communicate a simple overall message that would be understood by a lay audience. This short account explains the original scientific context for our research project, the sensitivities in dealing with a range of 'proxy' climate data and their interpretation, and the innovative outcome for dissemination that was devised in this case.

**Ideas to communicate**

While a short Press communiqué from a project can of course distil the science into a single idea that can be readily understood, it probably would not encapsulate the nuances of interpretation to which most practising scientists would aspire. Within our project there was a series of ideas that we wanted to communicate, including the following.

1. *The wider context for ACCROTELM:* The Intergovernmental Panel on Climate Change (IPCC), established in 1988, heralded an unprecedented growth in research and publications concerned with climate change (Fig. 1). The wider context for the ACCROTELM project was provided by the Third Assessment Report (Houghton et al. 2001), which as well as reinforcing the view previously expressed that the „balance of evidence suggests a discernible human influence upon global climate" (Houghton et al. 1996, p. 6), recognised scientific uncertainties. For example, the role of the sun in both past and recent climate change, and its incorporation in climate models (Stott et al. 2003), is uncertain.

Fig. 1: Results of combined keyword search of Web Of Science (WoS – now Web of Knowledge) Science Citation Index (SCI), Social Science Citation Index (SSCI) and Arts and Humanities Citation index (A&HCI) from 1981–2002. Dates of publications of the Inter-Governmental Panel on Climate Change (IPCC) also shown

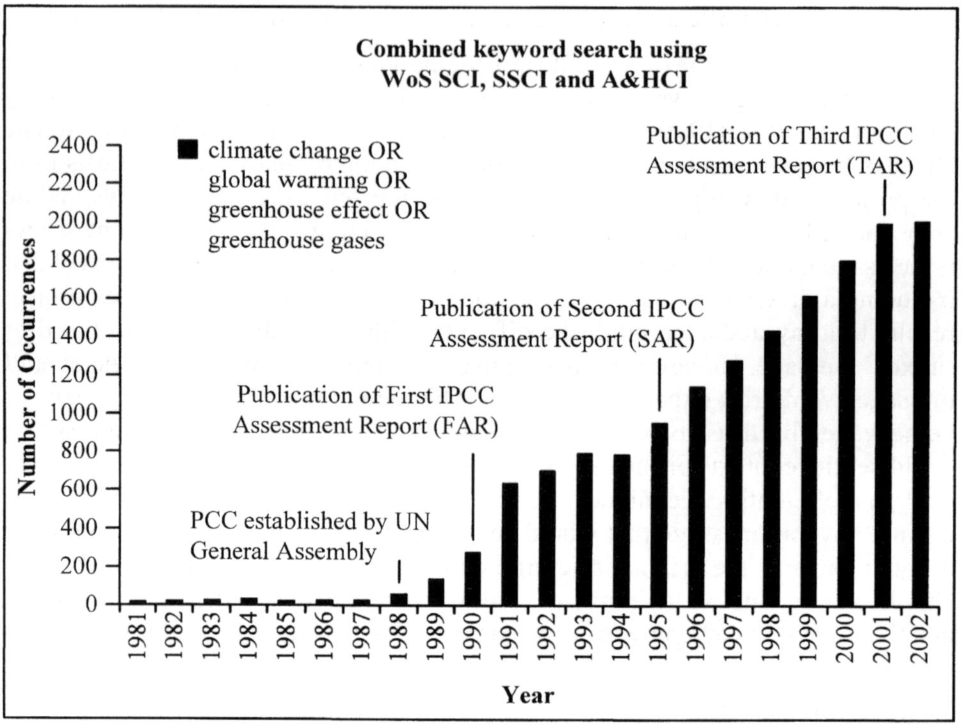


Chambers and Brain, 2002

2. *The discipline-specific context for ACCROTELM:* ACCROTELM grew from a concern among a group of European palaeoecologists that in recent scientific attempts to produce reconstructions of past climate, one archive that had been under-exploited was that of peat bogs (cf. Barber 1981; Chambers & Charman 2004), partly owing to some misconceptions as to how peat bogs grew (Backeus 1990).

3. *The value of palaeoclimate records:* Owing to the temporal limitations of the meteorological instrumental record, to ascertain the magnitude, rate and frequency of past climate change over past millennia requires the examination of 'proxy' records of past climate – from a range of natural archives such as ice cores, tree rings, peats, lakes, speleothems, etc. Attribution of the precise magnitude of the human-induced component to recent climate change is dependent upon reliably connecting the recent period of instrumented meteoro-

logical records to the proxy-climate records of earlier times, and on calibration between the instrumental and proxy-climate data (Oldfield 2005).

4. *The climate archives to be used in ACCROTELM:* In addition to new data from ombrotrophic peat bogs located on transects across Europe, the project involved comparison with data from changing lake levels recorded at lake sites in West-Central Europe (e.g. Magny 2004, 2007).
5. *The field and laboratory techniques used in the project:* Established techniques of determination of peat humification (Blackford & Chambers 1993) and of plant macrofossil analysis (Barber et al. 1994) were adopted with common protocols, and combined with analyses of testate amoebae (Charman et al. 2000; Charman et al., in press) to generate proxy-climate data from the peat sites. The possibility of obtaining a temperature signal from mires (by biomolecular analysis) was explored.
6. *The overall conclusions from the project:* The overall conclusions were (i) evidence for considerable spatial variability across Europe in terms of hydrological response to past (and by implication, future) climate change, and (ii) the need for more research on a range of climate archives from a range of sites, for testing and validating climate models, so as to improve the models' reliability for generating credible climate scenarios for the future.

It was recognised it would be difficult to encapsulate all of the above in a single Press communiqué.

**Dissemination by film and DVD**

At an ACCROTELM Project Meeting, held in Amsterdam in January 2006, the possibility of creating a DVD from the project was mooted. As the Co-ordinating institution, we began to explore this, but were daunted by the costs. We subsequently were able to find additional funding (from national sources) to facilitate filming and editing, with a view to making an educational resource for the teaching of undergraduates and postgraduates. The scheduling of a major climate conference – HOLIVAR2006 (2006) – due to take place in London in June of that year, provided an opportunity for filming interviews with several of the ACCROTELM personnel and of their displayed posters. This concentration of interviewees in one location considerably reduced the travel and hire costs of a film crew. We also seized the opportunity to conduct interviews with other world-renowned climate scientists and palaeoecologists, external to ACCROTELM, who were attending the conference. These included palaeoclimatologists Prof. Ray Bradley and Prof. Michael Mann, climate modeller Prof. Paul Valdes, and Quaternary environmental scientists Prof. Frank Oldfield – former Executive Director of PAGES (PAGES 2007), Prof. Rick Battarbee (Head of the Environmental Change Research Centre, University College, London), Prof. Neil Roberts,

Dr. Bernd Zolitschka, Dr. Juerg Beer, Prof. Danny McCarroll – Co-ordinator of an EC 6th Framework project 'MILLENNIUM', and Dr. Mark Maslin.

Fig. 2: Structure of the ACCROTELM Climate Change DVD variants (see text). Items below the dotted line are examples of possible additional material, intended for teaching purposes in Higher Education

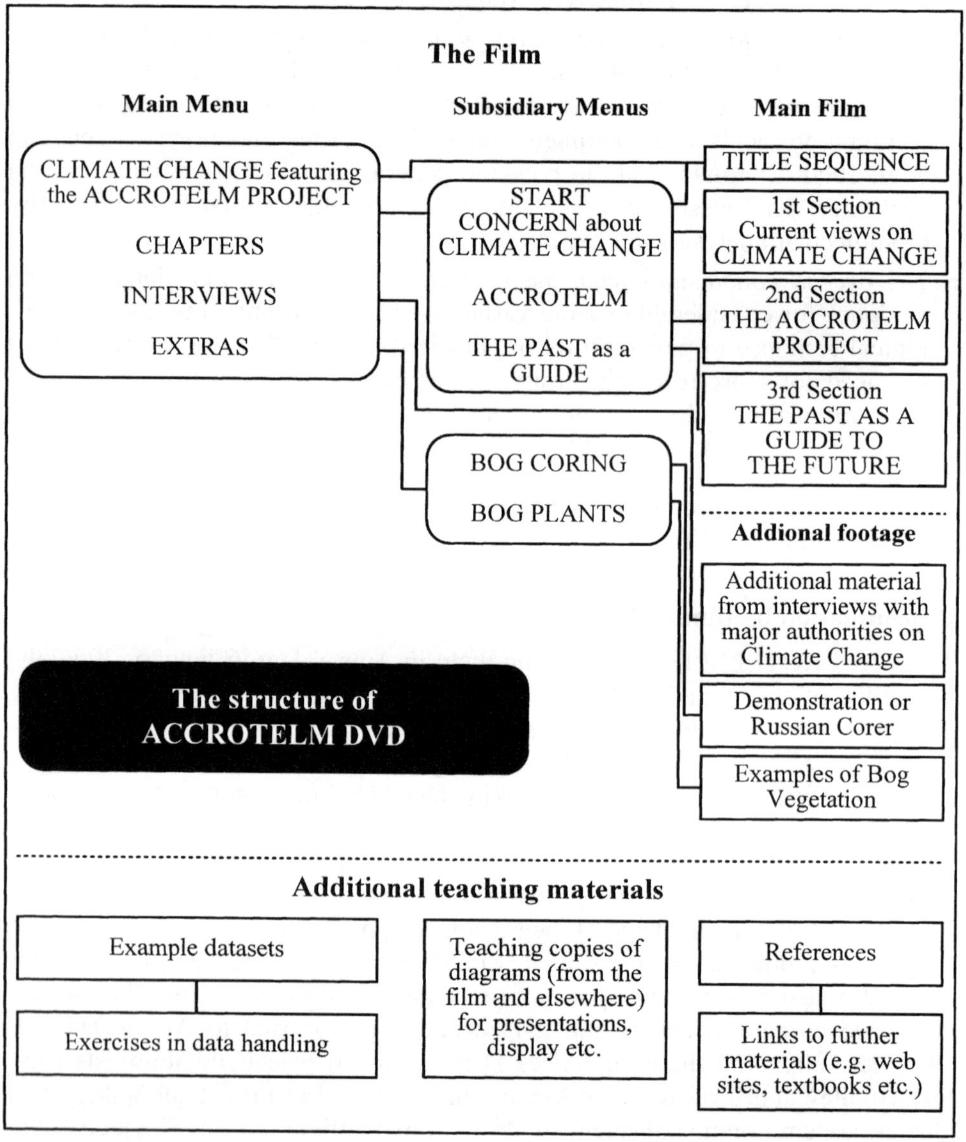


The resulting material is incorporated in three DVD versions (see Fig. 2). The first is a 35-minute film on Climate Change, featuring at its centre an account of the

project work, entitled '*Climate Change, featuring the ACCROTELM project*'. This is intended for wide distribution, including being made available to policy-makers and planners via the ESPACE 6$^{th}$ Framework project, and was incorporated in the Final Report of ACCROTELM. The film contains interview footage with leading palaeoclimatologists, to provide the scientific concerns, concepts and uncertainties; illustrates methods used in ACCROTELM to obtain 'proxy' records of the past climate of Europe; and emphasises the value of palaeoclimate data for testing and refining climate models. The film is permanently available on the Internet, hosted in low resolution by a Dutch TV/radio programme – Nooderlicht (2007) – so (at least potentially) ensuring wide dissemination.

A second variant of the DVD, produced with funds from a national teaching and learning project, includes additional interviews and supplementary material ('Extras' in Fig. 2). This variant has already been shown to Final-Year undergraduates at the co-ordinating institution, and at the University of Durham, UK. A third variant, yet to be produced, is proposed to contain links to additional resources, and possibly to datasets from the project, and is intended as an educational resource for undergraduates and postgraduates in Higher Education.

Despite evident imperfections, some of which could have been resolved with more funding and editing time, the film has generally been well received. However, we have learnt a great deal from the experience, and given the opportunity, we would approach any similar film/DVD venture somewhat differently next time.

## Acknowledgements

We thank Prof. Dr. Hansjörg Küster, for the original idea; ACCROTELM project members, for their time and contributions; Title Films, in particular Mary Addison and Liz Davies, for film and DVD production; interviewed scientists, for freely giving of their time; and HOLIVAR2006 organizing committee, for facilities. ACCROTELM contributes to Key Action „Global Change, Climate and Biodiversity", under the European Commission Fifth Framework Energy, Environment and Sustainable Development Programme, Contract no. EVK2-CT-2002–00166.

## References

ACCROTELM (2007a), „ACCROTELM (**A**brupt **C**limate **C**hanges **R**ecorded **O**ver **T**he **E**uropean **L**and **M**ass)", available at http:// www2.glos.ac.uk/accrotelm/Accessed 22/01/07.

– (2007b), „Publications1: Peer-reviewed articles", available at http://www2.glos.ac.uk/accrotelm/publications.html Accessed 22/01/07.

– (2007c), „Publications2: non-refereed literature", available at http:// www2.glos.ac.uk/accrotelm/publication2.html Accessed 22/01/07.

– (2007d), „Posters at HOLIVAR 2006", available at http://www2.glos.ac.uk/accrotelm/holipost.html Accessed 22/01/07.

*Backeus, I.* (1990), „The cyclic regeneration on bogs – a hypothesis that became an established truth", Striae, vol. 31, 33–35.

*Barber, K.E.* (1981), Peat Stratigraphy and Climatic Change, Balkema, Rotterdam.

*Blackford, J.J. & Chambers, F.M.* (1993), „Determining the degree of peat decomposition in peat-based palaeoclimatic studies", International Peat Journal, vol. 5, pp. 7–24.

*Barber, K.E., Chambers, F.M., Maddy, D., Stoneman, R. & Brew, J.* (1994), „A sensitive high-resolution record of late-Holocene climatic change from a raised bog in northern England", The Holocene, vol. 4, pp. 200–207.

*Chambers, F.M. & Brain, S.A.* (2002), „Paradigm shifts in Late-Holocene climatology?" The Holocene, vol. 12, 239–249.

*Chambers, F.M. & Charman, D.J.* (2004), „Holocene environmental change: contributions from the peatland archive", The Holocene, vol. 14, pp. 1–6.

*Charman, D.J., Hendon, D. & Woodland, W.* (2000), The Identification of Peatland Testate Amoebae, Quaternary Research Association Technical Guide no. 9, London. 147 pp.

*Charman, D.J., Blundell, A. & ACCROTELM Members* (in press), „A pan-European testate amoebae transfer function for palaeohydrological reconstruction on ombrotrophic peatlands", Journal of Quaternary Science, in press.

CORDIS (2007), „EUROPA – CORDIS: Community Research & Development Information Service", available at www.cordis.lu Accessed 22/01/07.

HOLIVAR2006 (2006), „HOLIVAR2006 Open Science Meeting Natural Climate Variability and Global Warming", available at www.holivar2006.org/ Accessed 22/01/07.

*Houghton, J.T., Meira Filho, L.G., Callander, B.A., Harris, N., Kattenberg, A. & Maskell, K.* (eds.) (1996), Climate Change 1995: the Science of Climate Change, Cambridge University Press, Cambridge.

*Houghton, J.T., Ding, Y., Griggs, D.J., Noguer, M., van der Linden P.J. & Xiaosu, D.* (eds.) (2001), Climate Change 2001: The Scientific Basis. Contribution of Working Group I to the Third Assessment Report of the Intergovernmental Panel on Climate Change, Cambridge University Press, Cambridge.

*Magny, M.* (2004), „Holocene climate variability as reflected by mid-European lake-level fluctuations and its probable impact on prehistoric human settlements", Quaternary International, vol. 113, pp. 65–79.

– (2007, in press), „Holocene fluctuations of lake levels in west-central Europe: methods of reconstruction, regional pattern, palaeoclimatic significance and forcing factors". In Elias, S.A. (ed.), Encyclopedia of Quaternary Science, Amsterdam, Elsevier, pp. 000–000.

*Mann, M.E., Bradley, R.S. & Hughes, M.K.* (1998), „Global-scale temperature patterns and climate forcing over the past six centuries", Nature, vol. 392, pp. 779–787.

– (1999), „Northern Hemisphere temperatures during the past millennium: inferences, uncertainties and limitations", Geophysical Research Letters, vol. 26, pp. 759–762.

Nooderlicht (2007), „Nooderlicht Nieuws. Nóg een klimaatfilm: Lessen voor de toekomst uit Europa's verleden", available at http:// noorderlicht.vpro.nl/ artikelen/31866284/Accessed 22/01/07.

*Oldfield, F.* (2005), Environmental Change: Key Issues and Alternative Approaches, Cambridge University Press, Cambridge.

PAGES (2007), „PAGES – Past Global Changes", available at www.pages.unibe.ch/Accessed 22/01/07.

*Stott, P.A., Jones, G.S. & Mitchell, J.F.B.* (2003), „Do models underestimate the solar contribution to recent climate change?", Journal of Climate, vol. 16, pp. 4079–4093.

## Chapter 12

# Raising Awareness of Climate Change: Experience in Lithuania

**Arunas Bukantis, Linas Kliucininkas, Egidijus Rimkus, Elena Talockaite**

## Abstract

Climate change is getting increased political attention all over the world. Academic research and studies, as well as formal and informal education are prominent participants in the public debate on climate change. Results of the public opinion poll show that although most people in Lithuania are aware of climate change and its threats, additional reliable information is highly required. Awareness-raising activities on climate change issues are foreseen in several national strategic documents. However, awareness-raising activities implemented in Lithuania in recent years could be considered to be isolated instances with a lack of monitoring of implementation and achieved results, lessons learnt, as well as a lack of continuity and coordination between different activities. New awareness-raising methods have to be created and used in order to attract as many target groups as possible.

This chapter gives an overview of the main outcomes of the survey of public opinion on climate change issues in Lithuania. Recommendations for awareness-raising in different target groups, as well as several good examples of awareness raising on climate change in engineering studies and activities carried out by non-governmental organizations are presented.

## Introduction

Seasons are changing, ice caps are melting, oceans are warming, plant and animal ranges are shifting – all because the climate is changing like never before. Global temperatures increased by about 0.6°C over the course of the last century, and are predicted to rise even more rapidly in coming decades. Most scientists argue that increased levels of carbon dioxide and other polluting gases in the atmosphere cause climate change (IPCC 2001).

The countries that are Parties to the UN Framework Convention on Climate Change have accepted certain commitments, taking into account their common but differentiated responsibilities and other specific national and regional development priorities. To fulfil these commitments, the Parties need to promote and facilitate various education, training and awareness-raising programs at the national and regional levels as required under Article 6 of the UN Framework Convention on Climate Change (United Nations 1992).

Climate change is getting increased political attention all over the world, but perhaps particularly in the European Union. Academic research and studies, as well as formal and informal education are prominent participants in the public debate on climate change.

The first and foremost step in regard to education, training and raising public raising is to develop a process nationally and regionally, to prepare teaching material to impart education, training and raise public awareness. This has to be done under a multi-tier approach. The second step is to build capacity to assess vulnerability to climate change and to evaluate options for mitigation and adaptation, to develop human and institutional capacities for the implementation of the UNFCCC and the Kyoto Protocol.

This Lithuanian model of Education, Training and Public Awareness on Climate Change will certainly yield the desired results if it is implemented and monitored appropriately at the state level, government level, primary, middle and high schools level, community level, and at the individual level.

## Level of public awareness of climate change in Lithuania

In the first half of 2006, researchers and students of the Department of Hydrology and Climatology of Vilnius University carried out an opinion survey of Lithuanian inhabitants. The questionnaire used for the interviews covered global warming and climate change issues as well as their threats. The main aim of the survey was to get a full-scale outlook regarding public awareness and attitude about the problem. Obtained results could be used for the preparation of a national education strategy on climate change.

500 Lithuanian inhabitants of both genders, different ages and education were interviewed. Interviews took place in urban and rural areas of different regions of the Republic. The distribution of respondents by gender, age, education and residence is very close to real statistical distributions of the Lithuanian population. For this reason it's very likely that the survey results precisely reflect the opinion of the Lithuanian population about existing problems related to climate change.

In order to understand how important the problem is from the respondents' point of view, interviewees were asked to evaluate different future threats for the world and for Lithuania separately (see table 1) using a seven grade scale. Analysis shows that on a global scale, respondents see the largest threats related to social problems (drug or alcohol addiction, increase of crime rate), environmental pollution and inefficient use of natural resources. Interestingly, respondents think that fundamentalism isn't a real future threat either for the world or for Lithuania. Although the largest threats to Lithuania are very similar to those already mentioned, the average grade for every item is significantly smaller.

Tab. 1: Evaluation of main future threats for the world and Lithuania

| Threats | Average grade | |
|---|---|---|
| | **World** | **Lithuania** |
| Terrorism | 5.4 | 3.7 |
| Military conflicts | 4.9 | 3.4 |
| Inefficient use of natural resources | 5.5 | 4.4 |
| Social problems | 5.7 | 5.3 |
| Global warming and climate change | **5.4** | **4.4** |
| Famine and drinking water shortage | 5.0 | 3.7 |
| Development of genetic engineering | 4.8 | 3.7 |
| Explosive rise of world population | 4.3 | 2.9 |
| Diseases (such as AIDS) and epidemics | 5.4 | 4.6 |
| Fundamentalism | 4.1 | 3.0 |
| Environmental pollution | 5.6 | 5.2 |

Global warming and climate change is quite an important issue too. This kind of risk was taken into consideration especially in the past few years when higher recurrences of extreme events (such as long term droughts, flash rains, and powerful storms) and increased damages had been observed. This topic is more important for young respondents rather than for the elder part of the population. On the other hand, in almost all cases, the average grade put down by respondents with university education (for climate change as well) is slightly lower. It is quite possible that this segment of the population feels safer from an economical and social point of view and external threats are less intimidating to them.

Most respondents suppose that they are well (17%) or moderately (63%) informed about climate change issues and only 20% of the interviewees think that climate change is a little known or unheard of problem (Fig. 1a). It is surprising that older and less educated males from rural territories consider themselves (probably mistakenly) as the best-informed part of the population. On the other hand, the same group is prevalent among the people who have no information about this problem. Meanwhile younger respondents with university degrees usually describe themselves as moderately informed. It can be concluded that higher education levels and more information of different kinds (from different sources) stimulate additional demand.

Fig. 1: Viewpoint of the respondents concerning the climate change issue (a) and the main sources of information about this problem (b)

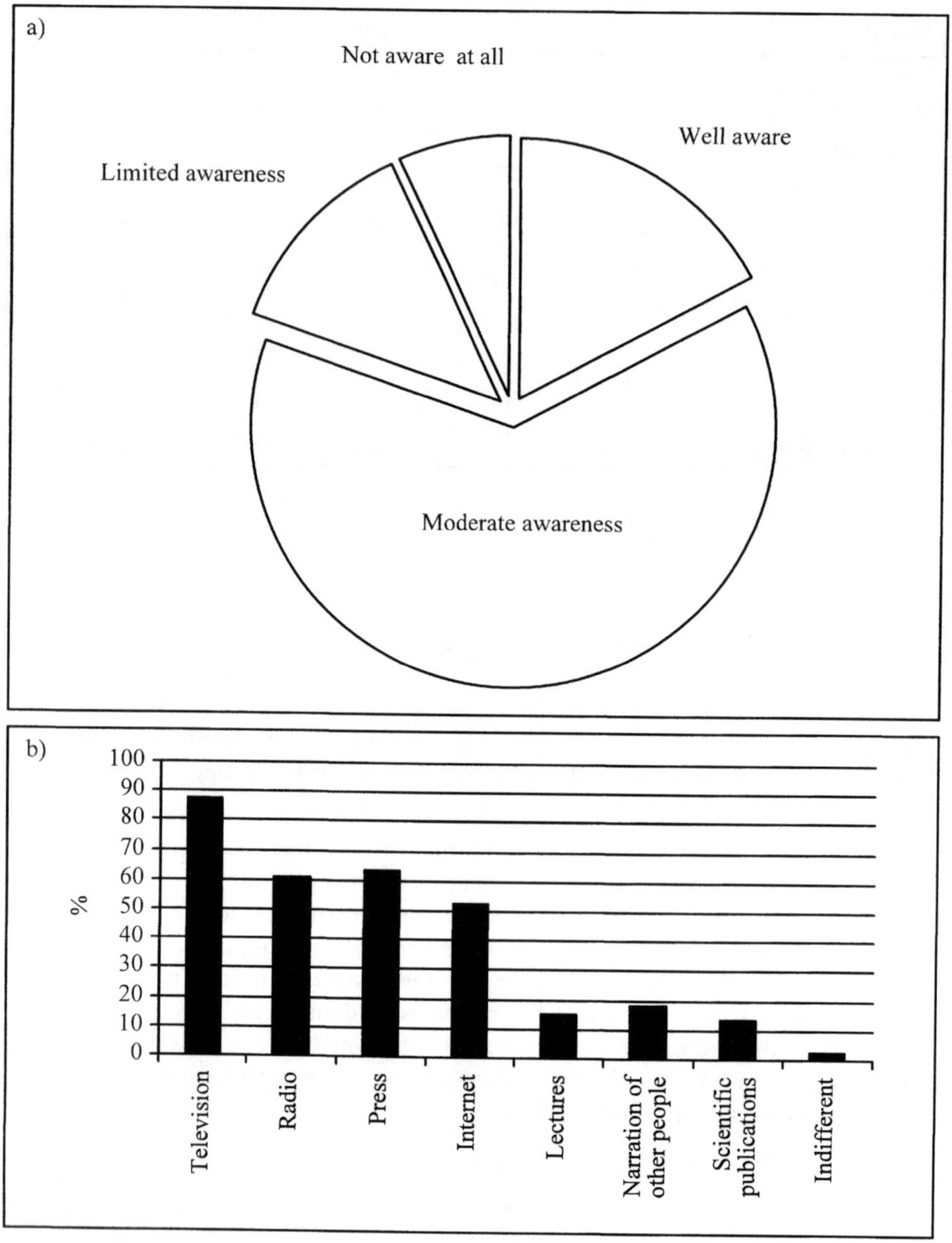


There are three main sources of information regarding climate change issues: television (87%), radio (61%) and the press (63%) (Fig. 1b). Although Internet is slightly behind (53%), this source of information is in second place (behind television) among the younger generation as well as part of the population with a university education. Without any doubt, the importance of the internet will largely increase in the future because of explosive growth of households having internet connections in the past few years, and the influence of this information source will be comparable with television. For this reason, one of the main strategic branches of raising awareness on climate change should be higher attention to the creation of web pages in national languages, as well as encouragement of scientists and environmentalists to prepare articles or short notes for internet pages.

Radio still remains a very important source of information for elderly people especially in rural territories, while the press is popular among middle-age urban populations. It is very important to tailor information to different social groups and to choose information sources correctly during awareness-raising campaigns.

Even 55% of respondents state that there is a lack of appropriate information about climate change issues. Most of them represent the well-educated middle-aged part of the Lithuanian population. It has to be noted that the majority of stakeholders representing different levels belongs to this social group. On the other hand, among the 38% of interviewees who claim that there is enough information in different mass media sources, the less educated elderly people prevail. The size of the completely indifferent population is quite small (7%).

The next two questions were related to the specific aspects of climate change. Firstly, respondents were asked to indicate the main associations that arise after climate change is mentioned. Most of them mentioned air temperature rise (65%) and glaciers melting (66%). The next important topics are sea level rise and an increase to the number and intensity of storms. The latter issues are especially important to the inhabitants of seashore zones. Secondly, respondents used a seven-grade scale to evaluate threats to Lithuania that arise from different aspects of climate change impacts. According to survey results, the greatest threats come from increasing droughts and more reoccurrence of flash rain. Representatives of rural populations also emphasized possible changes in the groundwater level. Usually their activities are very closely related to agriculture, which is probably one of the most vulnerable socioeconomic sectors. From the point of view of interviewees, there is very little threat arising from the increasing number of ecological refugees that could come to the EU or directly to Lithuania. Results show that our population does not realize the scale of globalization yet and can't clearly understand that although Lithuania isn't the most sensitive and vulnerable territory concerning climate change impact, extreme manifestations of changing climate, wherever it happens, will directly or indirectly effect social processes and the economic state of our Republic.

Fig 2: Evaluation of climate change threats for Lithuania

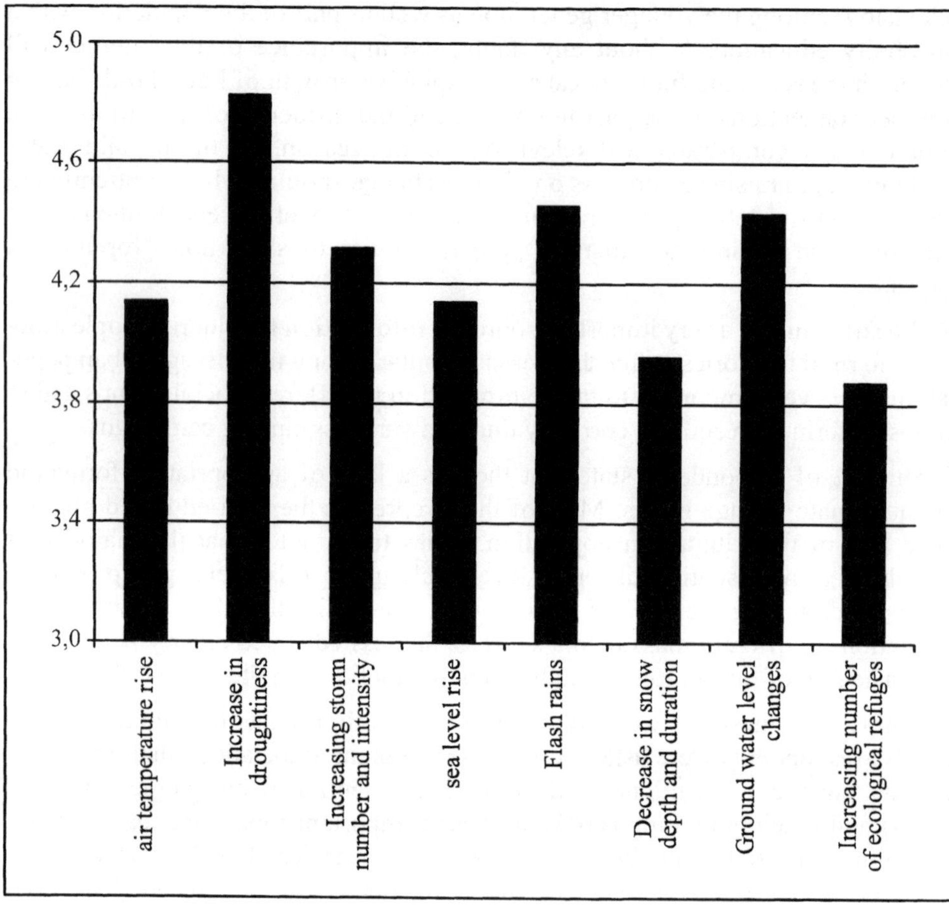


Although the largest proportion of respondents claim that they are well or moderate informed about climate change issues, answers to the next question relating to the Kyoto protocol show a slightly different situation. 87% of respondents couldn't correctly point out that this agreement is dedicated to climate change and greenhouse gas emission issues. Respondents with lower than secondary education, as well as the largest part (92%) of countrymen and, surprisingly, women were not able to answer. The number of correct answers didn't exceed 17% among people with university educations. Moreover, it is likely that this number is overestimated because of the possibility of implying the right answer from previous questions. Answers to this question clearly demonstrate the lack of information about climate change and efforts of the international community to find solutions.

On the other hand, 31% of the respondents agreed without reservation with a statement that Lithuania has to put in more effort and spend more money on the solution to climate change problems. Another 24% think that only money from the international foundations has to be allocated for this purpose. 28% of interviewees suppose that at national level these activities could be taken into consideration only after the economical situation in Lithuania has increased significantly. Other respondents claim that climate change is not an important problem at all. Elderly women are the most positive social group concerning such kinds of investments. Meanwhile, younger generations more often than others think that problems could be solved later.

44% of respondents suppose that in order to reduce greenhouse gas emissions at a national level, only measures that don't disturb the economical development of the country are suitable. Another 22% claim that such measures are unnecessary at all. Mostly well-educated middle-aged women are among the 34% of interviewees who accept even very expensive measures. In contrast, elderly and less educated persons assume that Lithuanian efforts in such activities are meaningless.

The last question was dedicated to the arguments used by skeptics. The respondents were asked to point out the most convincing arguments. 52% of respondents emphasized that that no one argument is convincing for them. 20% of respondents claimed that current climate change tendencies are only dependent on a natural cycle and human beings are not responsible. 15% of respondents assume that there is no problem in the near future and another 11% think that this issue is exaggerated. Skeptical arguments are more acceptable for older as well as less educated people. Meanwhile young and middle-aged university degree holders tend to approve of modern climate change theory.

It is possible to conclude that even if most people in Lithuania are aware of climate change and its threats, additional reliable information is required. Opinion surveys show that interviewees usually overestimate their knowledge on this topic. On the other hand, a lot of people have positive attitudes towards possible measures to mitigate and adapt to climate change at a local level. For this reason, this is a good time for political decision making. Some differences in the points of view of different target groups are visible. Therefore, these differences have to be taken into consideration while planning future awareness-raising campaigns. Greater attention has to be put on the selection of appropriate information and suitable information sources. Opinion surveys helped to identify the most important climate change issues from the point of view of Lithuanian inhabitants. These topics have to be emphasized in educational and climate awareness programs.

## Recommendations for awareness-raising activities

A roundtable discussion "Raising awareness of climate change" took place on February 6, 2006 in Vilnius (Lithuania) with a main aim to discuss and formulate strategic goals and actions for the education and awareness raising of different target groups on climate change issues. 17 participants in the event represented non-governmental organizations, education and science institutions, ministries, embassies and funds that support awareness-raising activities (ECAT 2006).

Participants of the roundtable discussion came up with a conclusion that awareness-raising activities on climate change issues are foreseen in several national strategic documents. However, there are not enough finances foreseen for the implementation of these activities. Lack of concrete persons or units allocated and responsible for the implementation of awareness-raising activities reduce the efficiency of the implementation process. Awareness-raising activities implemented in Lithuania in recent years could be considered to be isolated instances with a lack of monitoring of implementation and achieved results, lessons learnt, as well as a lack of continuity and coordination between different activities.

Governmental institutions carry out awareness-raising activities in the fields related to the climate change issues, such as promotion of sustainable mobility, alternative energy, etc. However, in many cases, the climate change aspect is not emphasized.

Within the framework of the project "Awareness-raising campaign on climate change" carried out by the Environmental Centre for Administration and Technology from December 2005 to May 2006 and financed by the British Embassy in Lithuania, recommendations for raising awareness on climate change were prepared (Bukantis 2006). Raising awareness on climate change aims at raising educated, independent, active and responsible members of the community, who would be aware of and be able to solve climate change related problems.

The main goals of the awareness raising process should be as follows:

- To form the community, which would be able to integrate into the European Union and actively participate in solving global climate and environmental problems.
- To inform target groups about climate change, possible impacts and the importance of adaptation in order to increase their ability to provide motivated criticism to the decision makers.
- To improve the dialogue between politicians, decision-makers, scientists, no-governmental organizations and communities at national, regional and local levels.
- To collect and present to the target groups information about good examples of solving climate change related problems in Lithuania and other countries.
- To stimulate integration of climate change issues to the strategic and spatial planning processes at local and regional level.

Recommendations include lists of measures for different target groups: preschool and schoolchildren; students of high schools and professional education institutions; local communities; architects, planners and environmental specialists; businessmen and farmers; politicians and other decision-making specialists of public authorities, etc. (see tab. 2)

Tab. 2: Recommendation of awareness-raising measures for different target groups

| Target group | Recommended measures |
|---|---|
| 1. Preschool and schoolchildren | 1.1 Formal education: integration of climate change issues into the existing curriculum of different subjects (biology, geography, chemistry, physics, mathematics, economics, art, etc.). |
| | 1.2 Informal education: lectures, discussions, development of the web pages in national language, competitions, interactive and computer games, etc. |
| | 1.3 Improvement of the environmental qualification of teachers, preparation of the teaching and information materials for the teachers. |
| | 1.4 Search for the lectors leaders among the teachers who would be able to distribute information about climate change. |
| 2. Students of high schools and professional education institutions | 2.1 Formal education: integration of climate change issues into the existing curriculum of different subjects. |
| | 2.2 Preparation of the teaching and video material for the separate course "Climate change". |
| | 2.3 Informal education: lectures, discussions, seminars, conferences, development of the web pages in national language, support for the activities stimulating public spirit of young people and their participation in the democracy processes, environmental protection activities. Raising solidarity and liberality of the youth. |
| | 2.4 Improvement of the environmental qualification of teachers. |
| | 2.5 Organization of events and publication of the information on climate change issues. |
| 3. Local communities | 3.1 Development of the system for public information about possible health impacts caused by the climate change and prevention of the new diseases (lectures, exhibitions and articles in media, specialized newspapers and magazines, TV and radio programs. |
| | 3.2 Education and information of public about healthy and environmentally friendly ways of living. |
| | 3.3 Development of the web pages in the national language in order to present important information about climate change and answer the questions raised by the public. |
| | 3.4 To stimulate national, regional and local media to play a bigger role in informing the public about climate change issues. |
| | 3.5 Organization of lectures and seminars for the representatives of the media in order to increase media interest in environmental problems and ensure a motivated and professional public information process. |
| | 3.6 Preparation of the individual environmental instructions concerning climate change, e.g., poster or sign in the office, pocket-sized card with behavior hints, checklist and recommendations for eco-driving. |

| Target group | Recommended measures |
|---|---|
| 4. Architects, planners and environmental specialists | 4.1 Lectures, seminars, conferences, discussions, qualification courses on climate change issues for specialists in construction, transport, coastal management, water management and landscaping.<br>4.2 Analysis of the new methods for solving climate change related problems based on scientific research and experience of other countries by organizing lectures, seminars and conferences.<br>4.3 Implementation of the pilot projects, monitoring of their implementation, results and lessons learnt.<br>4.4 Promotion of the idea of environmentally friendly houses, use of efficient energy technologies in construction, use of environmentally friendly construction materials and alternative energy sources. |
| 5. Businessmen and farmers | 5.1 Promotion of the scientific research and consultation concerning development and implementation of innovations and technologies in the field of greenhouse gases reduction, efficient use of energy and other natural resources.<br>5.2 Stimulation of international business exchange, participation in international research and new technology development programs.<br>5.3 Analysis of the impacts of climate change and possibilities for business by organizing practical seminars, conferences, feasibility studies, exhibitions, etc.<br>5.4 Implementation of pilot projects.<br>5.5 Stimulation of financial motivation (why it is worth caring about the environment?)<br>5.6 Stimulation and implementation of forest augmentation projects in order to increase absorption of carbon dioxide and regeneration of degrading land. Fostering the use of biomass and biofuel.<br>5.7 Comprehensive scientific research on the adaptation of agriculture and forestry to the climate change impacts. |
| 6. Politicians and other decision-making specialists of public authorities | 6.1 Awareness raising on climate change among the employees of governmental institutions and local authorities, information and education of elected politicians.<br>6.2 National support for the scientific research on climate change and use of research results while developing sectoral and environmental strategies. Broader implementation of climate change mitigation measures, assessment of the climate change impacts, etc.<br>6.3 Formation of the working group under the Ministry of Environment involving representatives of different institutions and target groups.<br>6.4 Financial support for the environmental monitoring system.<br>6.5 Cooperation and support for staff preparation for the education and consulting institutions, formation of public opinion on climate change issues.<br>6.6 Up-to-date and comprehensive public information about the quality of the environment.<br>6.7 Good example of climate change reducing behavior among governmental institutions, politicians, senior officials, i.e. energy saving in the offices, use of low-carbon means of transport, etc. |

Recommended measures have to be implemented using strategic and long-term thinking approaches in order to ensure good results and continuity of the awareness-raising process. It can be achieved by raising awareness of climate change strategy with clear goals and objectives, measures, roles and responsibilities involving different institutions.

## Some aspects of raising awareness of climate change in engineering studies

As a result of anthropogenic activities, emissions of carbon dioxide and other greenhouse gases during the last decade have increased considerably. The awareness of climate change could be perceived when studying different disciplines of engineering. This chapter provides examples of gaining knowledge on climate change through the energy, transport and industrial engineering studies.

### Reasons to reduce fossil fuel dependency

There are several arguments for sustainable energy systems. The reduction of greenhouse gas emissions by reducing fossil fuel dependency is one of them. The other is that the price of oil and gas will increase, even steeply, because the demand will not be met by production. The third concern is security. The fact that someone else controls the basic energy supply to a country is of course not secure and may, in the worst case, lead to conflict and blackmailing, and if not, costly negotiations if only one side in practice decides on the conditions (Aleklett 2006). This was illustrated by last year's events when Russia increased the price of gas delivered to Ukraine and Belarus.

Therefore understanding of current energy production, supply and usage regimes as well as principles of renewable energy production seem to be prime tasks for climate change education.

### The cost of transportation – energy and environmental impact

Transport energy amounts to some 20 percent of total final energy use in Europe. More than 90 percent of it is based on fossil fuels. Only in OECD countries, has the use of fossil fuels for transport increased by more than 45% from 1980 to 1997 and is expected to continue growing (OECD 2001). Alternative fuels in the form of non-crude oil resources are drawing considerable attention as a result of shrinking oil reserves, increasing petroleum costs and the need to reduce pollutant emissions. The emission of $CO_2$ is directly determined by fuel-use and the carbon content of fuels.

While the increase in total energy-use has been very moderate, energy-use for transport has increased significantly. Technical developments have reduced specific energy-use (MJ/vehicle-km) during the period, but this reduction has been offset by the very large increase in transport volume. A large fraction of the potential energy-efficiency improvements for vehicles has also been lost due to increases in vehicle weight and power.

There are also additional factors that have led to increased energy consumption. Firstly, transport has tended to use less energy-efficient modes of transport, such as train and air transport. Secondly, vehicles are used less efficiently. Finally, driving conditions, in particular the increasingly crowded roads, have also contributed to increasing energy use (Kliucininkas 2006).

Efficient energy use is heavily dependent on vehicle speed. Energy use is lowest at average speeds of approximately 60–80 km/h. At higher speeds, energy use increases rapidly. In cities it is the other way around. Here the speed is often much lower than optimal. Greater energy-consumption at low average speeds in urban traffic is a result of its inefficiency (Tengstrom & Thynell 1997).

## The unbalanced carbon cycle – industrial perspective

Human activities have substantially altered both pools and flows in the natural carbon cycle. The use of fossil carbon is around ten times greater than the total use of all metals in society. Flows of elements associated with the extraction of fossil fuels are actually greater than the amounts that are mined for several elements.

The preventive approach is widely implemented in industrial processes includes conserving raw materials and energy, eliminating toxic raw materials, and reducing the quantity and toxicity of all emissions and waste before they leave the process. The technological innovations are applied not only to industrial processes, but also to products and services. The strategy for products focuses on reducing impacts along the entire life cycle of the product, from raw materials extraction to the ultimate disposal of the product. Preventive approaches in services involve design issues, housekeeping improvement, and better selection of material inputs. The selection of what kind of service is to be provided and in what way leads to varying kinds of environmental impacts (Strahl 1997).

Fossil carbon is used to produce plastics and other organic chemicals, for example, lubrication oils, solvents and printing inks. When eventually broken down, they will also give rise to emissions of $CO_2$. In addition, some specific industrial processes give rise to net emissions of $CO_2$. Changing the production process may eliminate all greenhouse gases associated with certain industrial operations. There are branches of industries where the release of greenhouse gases is intrinsic to the product. In some cases, substitution of the materials is the only option for reduction of greenhouse gases. Cement production is one such example, where process-related emissions contribute 40 percent of total emissions of lithosphere $CO_2$ to the atmosphere (in calcination $CO_2$ is driven off from the carbonates). This means that they are much lower than energy-related emissions, but in the future global industrialized society, these emissions alone may reach a rate that is high enough to cause atmospheric concentrations to continue to increase even if emissions from the energy sector are phased out (Karlsson 1997).

The methodological approach discussed in this chapter was based on ten years experience of the Baltic University Program (BUP 2007). Very early BUP focused on sustainability issues. Problems related to climate change are the key issues discussed in the BUP network. The Program has created common educational resources relevant to interdisciplinary global environmental studies – a basic undergraduate course *A Sustainable Baltic Regio*n (1997), master level courses *Environmental Science for the Baltic Sea Basin* (2002) and *Environmental Management* (2006). The BUP courses do not constitute traditional academic studies. They instead address a problem, describe it, attempt to understand it, and seek for the solutions.

## Non-governmental organizations and international funds as a driving force

This section presents several examples of awareness-raising activities on climate change in Lithuania carried out by non-governmental organizations and main sources of financial support for such activities.

About 25% of all projects financed by the UNDP Global Environmental Fund Small Grants Program in Lithuania are related to climate change issues. The Program provides support for the projects that are initiated by local communities and non-governmental organizations and correspond to three main priorities: promotion of bicycles, alternative energy and energy efficiency. According to the program requirements, each project has to include an awareness raising element and project sustainability has to be ensured (ECAT 2006).

Experience of the Small Grants Program in Lithuania showed that there is a clear lack of climate change related projects due to the low public awareness and lack of knowledge of climate change and possible solutions in local communities. Interactive actions and information material (leaflets, articles in the newspapers, etc.) were named as the most efficient awareness-raising measures, while meetings with local communities organized in the framework of implemented projects were not so successful.

The British Embassy in Lithuania also pays a great deal of attention to climate change and raising public awareness. Since 2005, several projects on climate change issues have been financed. Representatives of the Embassy emphasized the lack of initiatives for such projects.

One of the projects financed by the British Embassy in Lithuania was carried out in 2005–2006 by the non-profit organization „Environmental Centre for Administration and Technology". Project "Awareness raising campaign on climate change" aimed at the raising awareness of Lithuanian society on climate change and development of possible actions of governmental and non-governmental sectors upon the problem. The project had the following objectives:

- to develop a strategic outline for education and raising awareness on climate change in Lithuania;
- to make different stakeholders groups in Lithuania aware of the climate change issue;
- to improve the dialogue between political level, decision-makers, scientists, NGOs and communities at national, regional and local levels;
- to gather and distribute to the stakeholders information about good practices on climate change mitigation, adaptation and awareness-raising that were implemented in Lithuania and other countries;
- to stimulate integration of climate change issues into the strategic and spatial planning processes at local and regional levels.

The objectives were reached by organizing national roundtable discussion, several regional events, creation of the database of good practices, an international conference, several publications in regional newspapers and dissemination of information about project activities and main outputs. Recommendations for raising the awareness of different target groups on climate change issues developed within the framework of this project are considered to be among the main results of the project. The document was distributed among relevant governmental and scientific institutions, as well as local authorities, local communities and NGOs. These recommendations are described in more details in previous sections.

The roundtable discussion organized within the framework of this project was one of the first events in Lithuania where representatives of different institutions and target groups discussed raising awareness on climate change. A dialogue between the political level, decision-makers, scientists, NGOs and communities was established.

In order to inform local target groups about the climate change issues, possible impacts on different areas (environmental, social, economical) and the importance of taking climate change into account while making strategic decisions and in the spatial planning process, several regional events were organized. More than 30 representatives of local and regional authorities (politicians/decision-makers, administration employees, planners), local NGOs, education institutions, and the media participated in each event.

The outcomes of these events are considered as successful mainly due to the right way chosen by the speakers to present the issue to the audience. The experience of Lithuania and other countries shows that scientists dealing with climate change issues carry out research activities and dispose of a lot of scientific data. It is essential to find a way to transfer this information to the public and decision-makers and to present it in a simple, attractive and understandable language.

Articles on climate change issues prepared by the specialists and published in regional newspapers gave a possibility to reach broader audiences. Experience with collection of information for the database of good practices was not very successful. Although there are a number of climate change mitigation and awareness-raising activities being carried out in different regions of Lithuania, responsible persons were not willing to provide descriptions of their activities and main outcomes. This could be explained by the lack of tradition of networking, promotion of the achievements and lessons learnt in Lithuania.

Experience of the Public Institution "Renewable energy information consultation center" shows that people in Lithuania don't see climate change as a problem. People are not willing to be frightened and to receive too much negative information. Therefore it would be more efficient to inform the public about possibilities to solve the problem, their possible role and hints for a change of behavior. Within the framework of several projects, the Center provided presentations on climate change issues during international and national events. The CD and information material on the possibilities of using solar energy in Lithuania was produced. This information is available in regional libraries. Implementation and marketing of the pilot projects could raise public awareness on climate change and stimulate the use of alternative energy.

Exhibitions, computer and interactive games, competitions, videoconferences and open discussions are considered to be the most efficient means of raising public awareness on climate change implemented in Lithuania by the British Council. The public is better at accepting new and interactive communication forms (ECAT 2006).

Non-governmental organizations play a significant role in the public awareness raising process on climate change issues in Lithuania. However, in most cases these activities are financed by different international funding sources, but not by national institutions. Therefore, it is quite complicated to ensure the continuity of awareness-raising activities.

## Conclusions

- In order to implement commitments to the international conventions, as well as requirements of strategic and legal documents related to climate change issues, it is essential to promote and facilitate various educational, training and awareness-raising programs at different levels. Academic research and studies, as well as formal and informal education are prominent participants in the public debate on climate change.

- It is essential to develop awareness-raising processes nationally, regionally and locally; to prepare teaching material to impart education, training and public awareness-raising processes; to build-up the capacity of different stakeholders to assess vulnerability to climate change and to evaluate options for mitigation and adaptation; to develop human and institutional capacities for the implementation of commitments and requirements related to climate change issues.
- According to the public opinion survey carried out in Lithuania in 2006, most respondents name social problems (i.e., drugs, alcohol, crime rate, etc.) as the greatest threats. Global warming and climate change is quite an important issue too. This can be explained by the increased number of extreme weather events and damage for the population in Lithuania in recent years. Most respondents describe themselves as being well or moderately informed about climate change issues. Television, radio and the press are named as main information sources. However the importance of the Internet is also increasing. A major proportion of the respondents didn't have a clear understanding of the Kyoto Protocol.
- Although most people in Lithuania are aware of climate change and its threats, additional reliable information is required. Opinion surveys show that interviewees usually overestimate their knowledge on this topic. On the other hand, a lot of people have positive attitudes towards possible measures to mitigate and adapt to climate change at a local level. For this reason, it is a good time for political decision making. Some differences in the points of view of different target groups are visible. Therefore, these differences have to be taken into consideration while planning future awareness-raising campaigns. More attention has to be put on the selection of appropriate information and suitable information sources. The opinion survey helped to identify the most important climate change issues from the point of view of Lithuanian inhabitants. These topics have to be emphasized in educational and climate awareness programs.
- Awareness-raising activities on climate change issues are foreseen in several national strategic documents. However, there are not enough finances foreseen for the implementation of these activities. A lack of specific persons or units allocated and responsible for the implementation of awareness-raising activities reduces the efficiency of the implementation process. Awareness-raising activities implemented in Lithuania in recent years could be considered as isolated instances with a lack of monitoring of implementation and achieved results, lessons learnt, as well as lack of continuity and coordination between different activities.
- Raising awareness on climate change should be considered in the context of sustainable development. Taking into consideration continuous changes in information about the climate and the development of new technologies, public knowledge has to be continuously updated. A possibility for life-long learning has to be provided. New awareness-raising methods have to be created and used in order to attract more target groups. Awareness-raising processes have to cover as many target groups as possible, e.g., public administration institutions (government, ministries, local authorities, etc.), education and scientific institutions, non-governmental organizations, business, communities, the media, etc.

- As a result of anthropogenic activities, emissions of carbon dioxide and other greenhouse gases during the last decade have increased considerably. The awareness of climate change could be perceived when studying different disciplines of engineering. Knowledge of climate change could be gained through the energy, transport and industrial engineering studies by the introduction of climate change issues to the university curricula, participation in relevant international programs, exchange of students, etc. Understanding the energy production, supply, usage regimes, principles of renewable energy and sustainable transport, as well as other engineering processes from the perspective of climate change has to become an essential element of engineering studies.
- The long-term efforts to establish interdisciplinary networks of professionals could be seen as a model for raising climate change awareness in academic society and society as the whole.
- Non-governmental organizations play a significant role in the public awareness raising process on climate change issues in Lithuania. However, in most cases these activities are financed by different international funding sources, but not by national institutions. Therefore, it is quite complicated to ensure the continuity of awareness-raising activities.

## References

*Aleklett, K.* (2006), Fossil fuels – how long will they last. Report from a Research Symposium on Sustainable Development Patterns. Baltic 21, ISBN: 91–975526–4–X, p. 75–80.

*Bukantis, A.* (2006), "Recommendations for Raising Awareness of Climate Change", Environmental Centre for Administration and Technology. Available www.ecat.lt.

ECAT (2006), "Raising Awareness of Climate Change", Minutes of the round-table discussion, Environmental Centre for Administration and Technology. Available www.ecat.lt.

IPCC (2001), "Climate Change 2001: The Scientific Basis. Contribution of Working Group I to the Third Assessment Report of the Intergovernmental Panel on Climate Change", Houghton, J.T., Ding, Y., Griggs, D.J., Noguer, M., van der Linden, P.J., Dai, X., Maskell, K. & Johnson, C.A. (eds.), Cambridge University Press, Cambridge, Unite Kingdom and New York, NY, USA, pp. 881. Available www.ipcc.ch/pub/reports.htm.

*Karlsson, S.* (1997), Man and material flows: towards sustainable materials management. The Baltic University Programme. ISBN 91–7005–126–7, p. 31–36.

*Kliucininkas, L.* (2006), Towards Environmentally Sustainable Transport. Report from a Research Symposium on Sustainable Development Patterns. Baltic 21, ISBN: 91–975526–4–X, 2006, p. 62–66.

*Kliucininkas, L., Zieniute, I. & Mockuviene, J.* (2005), An Assessment of the Co-generated Electrical Energy Potential for Lithuania. Management of Environmental Quality. Emerald, vol. 16 (4), p. 264–273.

Organisation for Economic Co-operation and Development (2001), OECD Environmental outlook. Paris, France: OECD Publications.

*Strahl, J.* (1997), Sustainable industrial production: waste minimization, cleaner technology and industrial ecology. The Baltic University Programme. ISBN 91–7005–128–3, p. 37–45.

*Tengstrom, E. & Thynell, M.* (1997), Towards sustainable mobility: Transporting people and goods in the Baltic Region. The Baltic University Programme. ISBN 91–7005–129–1, p. 13–18.

United Nations (1992), "United Nations Framework Convention on Climate Change". Available http://unfccc.int/ essential_background/ convention/ items/ 2627.php.

## Chapter 13

# Linking Sustainability, Education, Communication and Climate Change – Some International Approaches and Good Practice

**Walter Leal Filho, Franziska Mannke**


## Abstract

Sustainability is one of the key topics in the environment debate in the 21st century. However, despite the wide scope and diversity of initiatives focusing on sustainability, it is still regarded as an abstract topic among different groups. This indicates a perceived need to find ways to communicate it to different groups and audiences in a way that it allows it to be understood. This paper will describe the elements that need to be considered in bringing across the sustainability message to audiences in developing and industrialised countries and how to link it with the current debate on climate change. A special emphasis will be given to existing national and international networks and to project-related tools. Moreover, a sample of initiatives will be described, aimed at not only clarifying what sustainability is, but also the benefits that may be brought about by sustainable development communication in respect of climate change.


## Defining Sustainability

Sustainability is today one of the most widely used words in the scientific field as a whole.

It means different things to different people in different contexts (Baker, Kansis, Richardson, and Young, 1997) but is also seen as something related to better environmental quality (Blowers, 1993).

The analysis of the evolution of the concept of sustainable development is a rather difficult exercise (Leal Filho 2000). Until the late 1970s, the "word" sustainability was only occasionally employed in most cases to refer to ways through which forest resources should be used. At present, the expression "sustainability" is being used as a synonymous to words such as "long term", "durable", "sound" or "systematic" , among others. Indeed, out of the context of the English language, sustainable development is very often referred to as "durable development" in French, while word-by-word translations are found in the German (nachhaltige Entwicklung), Spanish (desarrolo sustenible) and Portuguese (desenvolvimento sustentável) languages.

A key question which one might ask at this stage -a question which is posed over and over again, every day, by millions of people all over the world – is *what does sustainable development really mean*? Depending on the ways it is looked at, it may have many meanings, such as:

- the systematic, long-term use of natural resources – as defined in the Brundtland Report described elsewhere in this chapter – so that these are available for future generations (here referring to country and local policies);
- the modality of development that enable countries to progress, economically and socially, without destroying their environmental resources (here referring to country policies);
- the type of development which is socially just, ethically acceptable, morally fair and economically sound (here referring to the social ramifications of development);
- the type of development where environmental indicators are as important as economic indicators (here referring to the close links it bears with economic growth).

Many other variants may be listed and are indeed used by different organisations, taking into account their political perspectives and institutional aims. IUCN, which in liaison with UNEP and WWF produced "Caring for the Earth" (IUCN, WWF, UNEP 1991), suggested at the time that the expression "sustainable development" be replaced in some contexts by "sustainable living", since although the suffix "development" is associated with governments and refers to a government's responsibilities, the word "living" is closer to an individual's life (Leal Filho 2000).

A significant step forward in the international efforts towards recognising the value and usefulness of sustainability, was the momentum generated by the publication of "Our Comon Future", also known as the "Brundtland Report" (WCED 1987), which reported on the deliberations of the *World Commission on Environment and Development* (WCED), an official group set up by the UN to investigate ways by which environmental conservation could be systematically pursued, internationally, paying due attention to economic, social and political considerations. After nearly three years of work, a panel composed by specialists from various disciplines and from different geographical regions provided an in-depth examination of what the matter of sustainability is, what is required to pursue it and the measures that countries should adopt so as to put into practice. Also, the definition of sustainability proposed by the WCED and referred to above, has contributed towards some degree of acceptance on what it means. This is an important achievement since, before that, emphasis to it was piecemeal.

## Linking sustainability and climate change

Sustainability may be applied to a wide range of situations and contexts, also in respect of climate change. Indeed, taking into account the global nature of climate change and the fact that ways to address it invariably lead to the engagement of key stakeholders and the general public, it is seen that a sustainability dimension is essential in the development of an action plan on climate change on a country, regional or global level.

But in order to link sustainability to climate change, it is important that one has an understanding of what it means to different people in different parts of the world. A recent study undertaken in the United Kingdom and disseminated by the British Broadcasting Corporation (BBC)*[1] has shown that climate change 'worries children'. A sample of young people was asked about the world's problems and climate change was often referred to as young people's biggest concern for the world's future. Some 24% of the sample believe it is the greatest threat faced, while 19% think it is crime and violence, the Department for Education and Skills found. Of the 1,000 10 to 18-year-olds questioned, 18% nominated terrorism and 12% said it was lack of housing around the globe which concerned them most.

In this context, it is important to acknowledge that there are some barriers seen when one tries to communicate climate change. One of them is related to some misconceptions of what climate change is. If not addressed, them may lead to a negative view of climate change.

Some of these misconceptions are:

1. climate change is too abstract a topic,
2. the theme climate change is too broad,
3. there are no trained people to handle it,
4. the amount of resources needed to explain what it is do not justify it,
5. the theme is polemic and has no reliable scientific basis.

When carefully examined, the above outlined misconceptions have quite deep roots and it is necessary to overcome them. Some of the factors that influence an individual's attitudes towards climate change are outlined in Table 1, which also show the importance of an efficient sustainable development communication strategy applied to climate change.

Four crucial elements need to be taken into account in order to promote education and raise awareness on climate change and hence allow it to be communicated more efficiently. These elements are:

- the need to motivate an increase in personal accountability;
- the need to engage all sectors of society;

---

1 Details at: http://news.bbc.co.uk/1/hi/education/4123884.stm.

- the need to document and disseminate successful experiences on education related to climate change and the merits of such educational efforts;
- the need to emphasis long-term perspectives as well as short-term implications of climate change.

It should be stressed that if citizens were disappointed in the past with the slowness with which traditional approaches to environmental protection have produced results, they are likely to face a similar problem in relation to climate changes: attempts to address it cannot be done immediately: it is a slow, long-term process which entails a broad education and building preparedness among individuals to be involved in the drive towards a more sustainable world. We should be aiming at not – only – more beautiful words, but to real commitment and engagement.

Tab. 1: Some factors which influence attitudes towards climate change

| | |
|---|---|
| **Knowledge** | Information on the meaning of climate change and its implications |
| **Background** | The nature of his/her training often influences an individual's degree of receptivity in relation to the topic of climate change |
| **Experience** | Previous experience with other environmental and social issues facilitate understanding on climate change and the role of sustainability |
| **Perception** | The integrated view of environmental, political and economic elements enable a broader perception of climate change and ways to address it |
| **Values** | Differing from the previous ones due to its high degree of complexity, an individual's values often determine whether his/her attitudes are favourable to climate change or negative to it |
| **Context** | Climate change is not only related to climate or ecological components per se, but also entail items such as economics, politics and social matters. However, links with the latter are often ignored |

Modified from Leal Filho 2000

In the successful communication of climate change and how it links with sustainable development, both children and adults must understand:

- how the natural world works;
- the interdependence of humans and the environment;
- the need to assess the effects on humans on the biosphere
- the need to analyse the links between climate change and human population dynamics; energy extraction, production and use; and other human activities such as agriculture, manufacturing, transportation, building and recreation;
- the relationship of population, consumption, culture, social equity and the environment;
- the interdependence of human health and the environment;
- the social, cultural, legal and governmental frameworks for guiding environmental management and sustainable development.

The above may only be reached via strategies targeted to motivate a more environmentally just and sustainable behaviour by both individuals and institutions.

In trying to establish a framework for promoting awareness on climate change, it may be said that climate change, as a theme, may be better understood by exploring the intersection of a number of different dimensions, such as the biophysical, cultural (including linguistic), economic, social, institutional, aesthetic and spiritual. This framework is not meant to be static; it must evolve as we improve our understanding of what climate change entails. The real world changes, and our understanding of it is full of uncertainty. Instead of only conveying facts, true education entails a process of interactive questioning and probing.

In a context as complex as climate change, students and adults alike should learn that systems thinking provides understanding rather than explanation, and that it emphasises:

- wholes over parts;
- relationships over objects;
- contextual over objective definitions;
- patterns over contents;
- quality over quantity;
- process over structure;
- dynamic equilibrium over stability;
- development over growth;
- inclusiveness over exclusiveness;
- non-linear dynamics; and
- complex cause-effect relationships.

Systems thinking is a mode of cognition that exists in us all, but tends to be de-emphasized and overshadowed by linear thinking within formal education. Failure to apply systems thinking when it is required often results in incomplete or erroneous solutions to problems.

This is unfortunately what happens often when one discusses problems related to climate change: the theme becomes detached from the day-to-day life and is seen as something caused by others. Systems thinking can lead to the understanding of the root causes of the various problems which are associated with and lead to climate change, hence catalysing more lasting, holistic and equitable solutions.

A framework designed to explain the approaches that may be used to communicate climate change is summarised in Figure 1. It brings the advantage of being adaptable to any environmental issue related to climate change and tries to take into account most variables that are seen in trying to bring it across to diverse groups and audiences.

Fig. 1: A framework to communicate climate change

| | |
|---|---|
| **Communicate what** | Climate change is a complex issue with many themes. Themes or topics thus need to be clarified and set of issues (e.g. energy, transport) have to be listed |
| | ⇩ |
| **Communicate to whom** | The target groups to be reached should be clearly outlined as recipients of the information (stakeholders) |
| | ⇩ |
| **How to communicate (approaches used)** | Talks, discussions, interviews, use of the media, briefings, reports |
| | ⇩ |
| **Dealing with scepticisms** | Economic data and ecological indicators are useful; Scepticisms may be neutralised with information on losses deriving from no action |
| | ⇩ |
| **Raising interest** | Bringing the global to the local level, with local examples and relating to one's responsibility; showing different contexts |
| | ⇩ |
| **Motivating action** | Emphasising the need for personal action, showing the consequences of inertia, illustrating – concretely – what can be done |

Not only factories and cars but also communities and institutions play a critical role in climate change and sustainable development. This is true for both industrialised and developing countries. In addition, both children and adults must understand the principles by which communities and institutions operate and influence climate change, so that they can also be shown how they can contribute to bringing about a sustainable future.

## Some International Initiatives

In Europe, there are a number of on-going initiatives in the field of climate change and sustainability. In the United Kingdom for example, the Environmental Change Network provides advice and disseminates information on various aspects of climate change (www.ecn.ac.uk/Education/climate_change.htm), including some tutorials which provide a sound basis for learning. The UK also hosts the "Atmosphere, Climate & Environment Information Programme", which disseminates information on the causes, effects of and controls on air pollution and climate change.

In the Netherlands, "The Climate Change and Biosphere Research Programme" (CCB-Wageningen) strengthens the cooperation between researchers in the field of fundamental, applied and policy-oriented climate change research. In France, the CERFACS (Global Change & Climate Modelling Project) undertakes simulations of the ocean-atmosphere climate and provide educational information on the topic. In Italy, the Institute of Atmospheric and Oceanic Sciences (ISAO) provides information on the dynamics and the physical, chemical, and biological processes of the atmosphere and of the ocean. In Denmark, the Climate Change Advisory Team of the Danish Energy Agency (DEA–CCAT) offers a full suite of models for calculating global GHG-scenarios and their effects on temperature rise and sea level change. In Germany, the "German Advisory Council on Global Change" (Wissenschaftlicher Beirat der Bundesregierung Globale Umweltveränderungen (WBGU)) considers its main task to be the evaluation of scientific knowledge concerning all aspects of global environmental change in the interdisciplinary context of the Earth system, and to formulate recommendations for political action. It also aims to promote greater awareness of global implications at all levels by means of education and public relations work.

The ASTRA project, already referred to elsewhere in this book, also has an educational dimension. The project, whose main objective is to assess regional impacts of the ongoing global change in climate and to develop strategies and policies for climate change adaptation, also assumes that there is a need to raise awareness among the various stakeholders so that reliable adaptation and mitigation strategies may be prepared. On its home page (Figure 1), users already have some insights into the project and its strong information dissemination nature.

Fig. 1: ASTRA's Home page

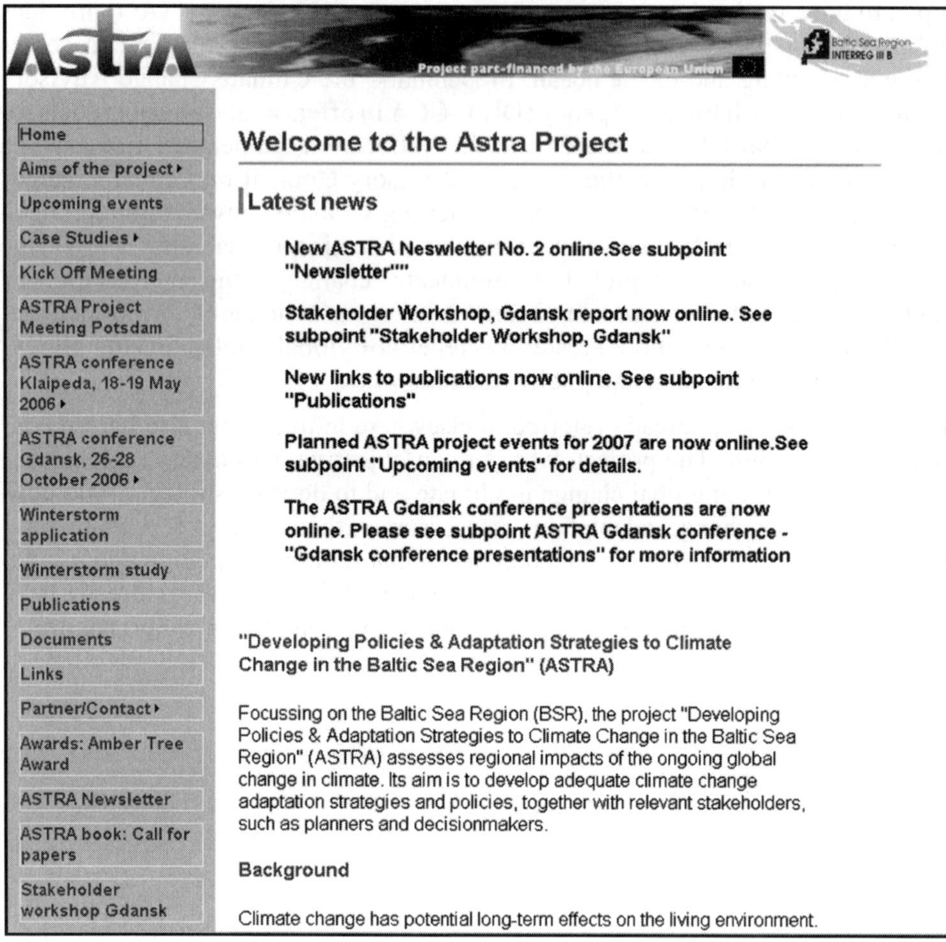


http://www.astra-project.org

The approach used in ASTRA is inclusive, i.e. it tries as much as possible to involve regional and local spatial planners and other stakeholders. It does so by identifying entry points via which integration to existing planning processes and methods are possible. Moreover, climate change impacts and vulnerability of regions are studied in several regional and local case studies jointly identified by scientists in cooperation with regional planners, in the following areas:

- Estonia: Estonian archipelago, Cities of Tallinn and Pärnu
- Finland: Pirkanmaa Regions, cities of Espoo, Raahe and Kokkola
- Germany: Oder River Estuary
- Lithuania: Curonian Spit, City of Klaipeda

- Latvia: Salaca River Basin, City of Riga
- Poland: Odra River Estuary, Cities of Gdansk and Sopot

The case studies add a strong local (and specific) dimension to the problems being assessed.

Out of Europe, some interesting educational initiatives are being performed, such as:

- The Climate Change Education Centre in Canada (http://www.conservationcorps.nf.ca/climatechange/html/links.html)
- The Australian Climate Change Education Resources (www.greenhouse.gov.au/education/index.html)
- The Atmospheric Radiation Measurement (ARM) in the United States, a programme involved in research and education with regard to the understanding of global climate change (www.uaf.edu/olcg/globalchange/index.html)
- The Climate Change Education home page contains useful information and links (http://climatechangeeducation.org/)

These examples, a few from a long list of initiatives, illustrate the wide range of information available and support the learning process among young people and adults alike.

## Conclusions

A long term view of climate change linked with sustainability requires the integration of community involvement, stakeholder identification, inter-institutional collaboration, and communication. The foundation of sustainable human development includes more equitable economic growth, poverty alleviation and a greater participation of people in making the decisions that affect their lives, and maintenance of environmental and ecological integrity.

It is important that these elements are considered in the current climate change debate.

Practical projects on climate change, such as ASTRA, can and should make more use of education and communication approaches, so as to inform the various stakeholders in a more meaningful way. To this purpose, an increased use of information technologies can support practitioners, government officials, academics and other groups in both planning and implementation of educational initiatives on climate change. This is potentially very useful in the sense that it can add efficiency in handling not only problems associated with climate change, but also support other sustainable development efforts.

## References

*Baker, S., Kansis, M., Richardson, D. & Young, S.* (eds.) (1997), The Politics of Sustainable Development. Routledge, London.

*Blowers, A.* (ed.) (1993), Planning for a Sustainable Environment. Earthscan, London.

IUCN, WWF, UNEP (1991), Caring for the Earth. Oxford University Press, Oxford.

*Leal Filho, W.* (ed.) (2000), *Communicating Sustainability.* Verlag Peter Lang, Frankfurt.

United Nations (1992), The UN Conference on Environment and Development: a Guide to Agenda 21. UN Publications Service, Geneva.

World Commission on Environment and Development/WCED (1987), Our Common Future. Oxford University Press, Oxford.

## About the Editors

Prof. *Walter Leal Filho*, Doctor of Philosophy (Ph.D.), Doctor of Sciences (D.Sc.), Doctor of Letters (D.L.) teaches sustainable development and environmental management in Hamburg, Stockholm and Naples, also being honorary professor at the A.I. Open University, Hochschule Zittau-Görlitz and the South-Western University "Neofit Rilsky" (SWU). He is the founding editor of the „International Journal of Sustainability in Higher Education" (launched in 2000) and of the journal "Environment and Sustainable Development" (launched in 2002), also being editor of the "European Environmental Education Newsletter" (launched in 1994). He has authored, edited or co-edited around 40 books on the subject matters of environmental education, environmental communication and sustainability, including "Introducing Concepts of Sustainable Development at University Level" (1996), "Sustainability and University Life" (1998 with a second volume in 2000), "Communicating Sustainability" (2000), "Teaching Sustainability at Universities: towards curriculum greening" (2002), "Integrative Approaches Towards Sustainability" (2004), "Handbook of Sustainability Research" (2005) and has in excess of 120 published papers to his credit.

TuTech Innovation GmbH
Life Sciences
Harburger Schlossstrasse 6-12
21079 Hamburg, Germany
www.tutech.de
email: leal@tutech.de

*Franziska Mannke*, B.A. in International Business Administration from the University of Southern Denmark and the University of Flensburg, earned her master's degree in International Business Administration from the University of Hamburg, department of Economics and Politics. She is currently coordinating the EU project ASTRA on the development of policies and adaptation strategies to Climate Change in the Baltic Sea Region and the EU project LifeSTYLE on renewable energy and sustainable technologies at TuTech Innovation.

TuTech Innovation GmbH
Department of Life Sciences
Franziska Mannke
Harburger Schlossstr. 6–12
D-21079 Hamburg
www.tutech.de, www.lifestyle-project.eu,
www.erneuerbare-energien-hamburg.de
email: mannke@tutech.de

*Philipp Schmidt-Thomé* works for the Geological Survey of Finland (GTK) as a researcher and project manager in European and international projects since

1998. He has been educated as a geographer (MSc.) an in applied environmental sciences, and holds a PhD in geology. In the last four years he has mainly focussed on natural hazards and climate change adaptation; focussing on stakeholder communication processes in regional development to support decision making processes.

Philipp Schmidt-Thomé
MSc. Geography, PhD. Geology
Senior Scientist
Geological Survey of Finland
Betonimiehenkuja 4
02150 Espoo
Finland
email: philipp.Schmidt-Thomé@gtk.fi

## About the Authors

*Eugenia Aloj* is full professor in Ecology in Sannio University (Faculty of Economics) in Benevento and professor in Environmental Education and Communication in Suor Orsola Benincasa University (Faculty of Training Sciences). She is also Chief of the Operative Unit „Biology and Ecology for protection of Environment" in Regional Centre of Competence *Benecon* in Campania Region. Since many years she has been involved in environmental education as education to sustainable development particpating in national and international projects on the topics. Her guidelines of research deal with environmental education to sustainable consumption (energy, transports, waste, agriculture) and with environmental education and environmental communication tools for companies and their corporate social responsibility. Mariagrazia De Castro and Anna Zollo worl with her participating to the didactic and reserach activities.

Prof. Eugenia Aloj
Sannio University
Via Nicola Calandra
82100 Benevento
Italy
www.eugeniaaloj.it

*Sally Brain* is a Research Associate of the University of Gloucestershire, having previously worked in the Coal Research Establishment, in England. Sally studied in Cheltenham for an MSc in Environmental Policy and Management, and for her MSc dissertation she analysed a peat core from Tierra del Fuego that showed evidence for an abrupt climate change 2800 years ago, coincident with climate changes recorded in Europe and elsewhere. She is currently researching solar influences on past climate.

*Arunas Bukantis*, Prof. Dr., is the Head of Department of Hydrology and Climatology at Vilnius University. His main research interests are the indices of climate change in Lithuania, historical records of thermal parameters, extreme climatic phenomena, climatic resources dynamics and forecasting, as well as regional anthropogenic climate changes and climate change impacts. Prof. Bukantis is involved in the development of national strategic documents on climate change issues and participates actively in awareness raising activities for various target groups.

Prof. Dr. Arunas Bukantis
Department of Hydrology and Climatology
Vilnius University
M.K.Ciurlionio 21/27
LT-01513 Vilnius
Lithuania
email: arunas.bukantis@gf.vu.lt

*Mariagrazia De Castro* – Learner in Ecology in in Sannio University

*Frank Chambers* is Professor of Physical Geography at the University of Gloucestershire, where he heads the Centre for Environmental Change and Quaternary Research (CECQR). Frank is interested particularly in the past record of climate and the forcing factors that drive natural climate changes. One of the principal climate archives he investigates is that of peat bogs: their stratigraphy, microfossils and plant macrofossils. A complementary interest is that of the vegetation history of bogs and moorlands, whether they have been influenced by human activity and by how much, and how they might be managed and conserved.

Prof. Frank M. Chambers
Centre for Environmental Change and Quaternary Research
Department of Natural & Social Sciences
University of Gloucestershire
Francis Close Hall
Swindon Rd
Cheltenham
GL50 4AZ
UK
email: fchambers@glos.ac.uk

*Remigijus Ciegis* – Doctor Habil. Professor, Kaunas Faculty of Humanities, Vilnius University. A graduate of Kaunas University of Technology as engineer-economist (1982) and Vilnius University (1989) as economist. In 1995, Doctor (VU, economics,), in 2002 Doctor Habil. (VDU, management). Author of more than 100 scientific papers and 8 monographs in Lithuanian and English. Member of International Society of Ecological Economics, member of editorial boards of several research journals. R. Ciegis has experience in studies related to soil, water, air monitoring, environmental planning, environmental policy and strategy, environmental economics, environmental and legal regulation. His research interests also include environmental management and economic methods of environmental regulation.

*Shane Conway* has an diploma in building construction from the University of Northampton, and has worked in building maintenance, building surveying and built environment. He currently works for Northampton Borough Council, England as an Energy Efficiency Officer.

Shane Conway
Private Sector Housing
Northampton Borough Council
Northamptonshire NN4 7NR
United Kingdom
email: sconway@northamton.gov.uk

*Jill Cook*, 'ESPACE – a new transnational approach to adapting to climate change' was edited by Jill Cook from Hampshire County Council. Jill has spent the last eight years working in the field of communications. Prior to ESPACE, she worked for the UK Environment Agency on a variety of national and European funded environmental projects. She has also worked for BBC Science Television, making programmes about the weather and climate.

Jill Cook
ESPACE Project
Hampshire County Council
The Environment Department
The Castle
Winchester
SO23 8UD
UK
www.espace-project.org

Contributing authors of this paper represent a number of organisations that collectively form the 'ESPACE Partnership'. This includes:

Hampshire County Council – Lead Partner (UK),
Environment Agency (UK)
Regionaal Landschap Zenne, Zuun en Zoniën – (Belgium)
South East Climate Change Partnership (UK)
South East England Regional Assembly (UK)
Surrey County Council (UK)
Waterschap Rivierenland – (the Netherlands)
West Sussex County Council (UK)
Ministerie van VROM – (the Netherlands)
Bayerisches Landesamt für Umwelt (Germany).

*John Daniell* is a Research Associate of the University of Gloucestershire. His original training was in Drama at the University of Bristol, and he subsequently had a career with the Greenwich Young People's Theatre. In later life John took an MSc in Advanced Ecology at the University of Durham, followed by a PhD studying the Late-Holocene Palaeoecology of Scots Pine in north-west Scotland. John has been Research Fellow at the University of Gloucestershire on two European Commission RTD projects: TIMECHS and ACCROTELM, and has worked on contract research investigating recent vegetation change in bogs and moorland.

*Birte Frommer*, Dipl.-Geogr., Degree in Geography at the J.-W. Goethe Universitaet Frankfurt am Main/Germany in spring 2000. From then on to 12/2002 environmental consultant at Infrastruktur und Umwelt/ Darmstadt, engaged in projects on environmental management and sustainable development at local and regional level, project management and GIS. Since 01/2003 assistant researcher and lecturer at the Section Environmental and Spatial Planning.

Research projects at the TU Darmstadt:

- KLARA-NET – Network on climate change adaptation in the Region of Starkenburg (Funded by the Federal Ministry of Education and Research (BMBF), 2006–2007)
- Moderation and specialist advice for the Regional Agenda 21 Starkenburg (1998–2005)

- Midterm Evaluation Regional Agenda 21 Starkenburg (2005)

Dipl.-Geogr. Birte Frommer
Technische Universität Darmstadt
Institut WAR – Fachgebiet Umwelt- und Raumplanung
Petersenstraße 13
64287 Darmstadt
email: b.frommer@iwar.tu-darmstadt.de

*Dalia Gineitiene* – Junior Researcher, Socio-Cultural Research Centre, Kaunas Faculty of Humanities, Vilnius University. A graduate of Vilnius University (1982) and Kaunas University of Technology (1996). Master in Public Administration. Research interests: sustainable development, sustainable development management, education for sustainable development, public participation.

Dalia Gineitiene
Kaunas Faculty of Humanities
Vilnius University
Muitines str. 8
LT-44280 Kaunas
email: Dalia.Gineitiene@vukhf.lt

*Philip Gray* has an M.Sc. in Environmental Technology from Imperial College, London and has worked as a researcher in sustainable development, risk management and communication about risks in Norwich, England and Jülich, Germany. He is currently the Environmental Performance Advisor for Northamptonshire County Council, England.

Philip Gray
Property Services
Northamptonshire County Council
Northamptonshire NN1 1AS
United Kingdom
email: pcgray@northamptonshire.gov.uk

*Simo Haanpää*, M.Sc. (Human Geography), works as researcher at the Centre for Urban and Regional studies (YTK) at the Helsinki University of Technology. A researcher at YTK since 2004, Mr. Haanpää is working on national and international projects dealing with climate change related questions in spatial planning. Past projects include an overview on the consideration of flood issues in spatial planning in Finland (EXTREFLOOD) and contribution to the study on adaptation needs in Finland from spatial planning point of view (FINADAPT). Currently he is working for an Interreg IIIB Baltic Sea –project ASTRA (Developing Policies & Adaptation Strategies to Climate Change in the Baltic Sea Region).

Simo Haanpää
Researcher, M.Sc.
Helsinki University of Technology
YTK Centre for Urban and Regional Studies
email: simo.haanpaa@tkk.fi

*Lena Herlitzius*, Dipl.-Ing., degree in Spatial Planning at the Universitaet Dortmund in autumn 2004. Traineeship at the Institut für Kommunikation und

Umweltplanung (iku GmbH) in Dortmund from October 2004 to April 2005. Assistant researcher and lecturer at the Section Environmental and Spatial Planning since April 2005.

Research projects at the TU Darmstadt

- KLARA-NET – Network on climate change adaptation in the Region of Starkenburg (Funded by the Federal Ministry of Education and Research (BMBF), 2006–2007)
- SRDTOOLS – Methods and Tools for evaluating the impact of Cohesion Policies to Sustainable Regional Development (6. EU-Framework Programme, 2005–2006)
- „Freude am Fluss" – Co-operative planning process in flood prevention (INTERREG IIIB-project, 2003–2008)

Shared responsibility on flood protection; pilot project of the Region Starkenburg for INTERREG-IIIB „Freude am Fluss"

Dipl.-Ing. Lena Herlitzius
Technische Universität Darmstadt
Institut WAR – Fachgebiet Umwelt- und Raumplanung
Petersenstraße 13
64287 Darmstadt
email: l.herlitzius@iwar.tu-darmstadt.de

*Hans-Jörg Isemer* holds a PhD in meteorology, his main research interests have been in the areas of air-sea exchange and the exploitation of ship observations for climate research purposes. He is leading author of a climate air-sea interaction atlas of the North Atlantic Ocean. Since 10 years he is head of the International BALTEX Secretariat, where BALTEX (Baltic Sea Experiment, see www.baltex-research.eu) is a regional hydro-climate programme within GEWEX and WCRP.

Dr. Hans-Jörg Isemer
GKSS Forschungszentrum Geesthacht GmbH
International BALTEX Secretariat
Max-Planck-Straße 1
D-21502 Geesthacht
Germany
email: isemer@gkss.de
www.baltex-research.eu

*Johannes Klein* has worked for the Geological Survey of Finland since 2004. He graduated from the University of Stuttgart in environmental engineering. His current research focus is on risk assessment, climate change adaptation and regional and urban development. He has participated in EU-wide projects in the fields of GIS, hazard and risk assessment and spatial planning.

Johannes Klein
Research Scientist
Geological Survey of Finland
Betonimiehenkuja 4
02150 Espoo

Finland
email:johannes.klein@gtk.fi

Prof. *Linas Kliucininkas* is the Head of the Department for Environmental Engineering at Kaunas University of Technology, Lithuania. He has extensive record of publications covering fields of environmental and sustainability science. A special interest is paid on monitoring and modelling, integrated assessment of the environmental impacts from transport and energy sectors, as well as local indicators of the sustainable development. The major part of the publications is of an applied character and has an interdisciplinary nature. During last decade Kliucininkas has initiated and participated in a number of international programmes and projects supporting environmental and sustainability studies in the Baltic Sea Region.

Department for Environmental Engineering
Kaunas University of Technology
Radvilenu 19,
LT-50254 Kaunas
Lithuania
email: linas.kliucininkas@ktu.lt

*Evangelos I. Manolas* was born in Naxos, Greece, in 1961. He has received a Bachelor of Arts in Sociology from the University of Essex (1983), a Master of Arts in International Relations from the University of Kent at Canterbury (1985) and a Philosophy Doctorate from the University of Aberdeen (1989). He is Assistant Professor of Sociology and Environmental-Forest Education in the Department of Forestry and Management of the Environment and Natural Resources at the Democritus University of Thrace

Evangelos I. Manolas, BA, MA, PhD
Assistant Professor
Department of Forestry and Management of the Environment and Natural Resources
Democritus University of Thrace
193 Pantazidou Street
68200 Orestiada
Greece
email: emanolas@fmenr.duth.gr

Prof. *Anders Omstedt* has participated actively in several national and international scientific bodies and has an active role presently in BALTEX (http://w3.gkss.de/baltex/, the MUSCAD (www.geol.lu.se/proxy/) and the EUR-OCEANS (www.eur-oceans.org/) programs. His main work has been oriented towards the ocean climate and the physics of ice covered water bodies, in particularly the Baltic Sea and the polar oceans. He has developed several numerical models for cold water bodies such as rivers, lakes, estuaries, shelf seas and the polar oceans. He is Depute head of Department of Earth Sciences, Göteborg University, Head of Ocean Climate Group at Göteborg University, Earth Sciences Centre-Oceanography and vice-chair men in BALTEX. A full CV with publications can be found on www.oceanclimate.se.

Prof. Anders Omstedt
Göteborg University
Earth Sciences Centre, Box 460
SE-405 30 Göteborg, Sweden
email: anders.omstedt@gvc.gu.se
www.baltex-research.eu

*Lasse Peltonen*, D.Adm.Sc. (Environmental policy), works as senior researcher at the Centre for Urban and Regional studies (YTK) at the Helsinki University of Technology. His research themes include environmental conflicts, environmental movements, governance, and public participation. During the past three years, Lasse Peltonen has worked in projects concerned with the linkages between risk governance and planning practices – with a special focus on climate change adaptation. Currently he is working for an Interreg IIIB Baltic Sea – project ASTRA (Developing Policies & Adaptation Strategies to Climate Change in the Baltic Sea Region).

Lasse Peltonen
Senior researcher, Dr.Adm.Sc.
Centre for Urban and Regional Studies
Helsinki University of Technology
P.O.Box 9300, 02015 TKK
Finland
email: lasse.peltonen@tkk.fi
www.tkk.fi/Units/Separate/YTK/index.html

Dr. *Egidijus Rimkus*, Associated Professor at the Department of Hydrology and Climatology at Vilnius University. His research interests are investigation of climate change and evaluation of its impact on nature and social processes in Lithuania, extreme climatic phenomena, dynamics and forecasting of climatic resources. He actively took part in several national and international projects related with different aspects of climate change. Dr. Rimkus is also involved in preparation of National Strategy and Action Plans for the implementation of the United Nations Framework Convention on Climate Change. He published over 40 scientific papers.

Department of Hydrology and Climatology
Vilnius University
M.K. Ciurlionio 21/27
LT-03101 Vilnius
Lithuania
email: egidijus.rimkus@gf.vu.lt

Privat-Dozent Dr. habil. *Gerald Schernewski* – Senior scientist and head of the working group „Coastal Research and Management" at the Baltic Sea Research Institute Warnemünde (Leibniz-Institut für Ostseeforschung Warnemuende). Associate lecturer at the Universities of Kiel and of Rostock and guest-professor at the University of the Algarve, Portugal as well as the St. Petersburg State University, Russia. President of EUCC – The Coastal Union Germany and member of the Executive Committee of EUCC – The Coastal Union International.

Prof. *Hans von Storch* is director of Institute of Coastal Research of the GKSS Research Centre and professor at the Meteorological Institute of the University of Hamburg. From 1987–1995, he was Senior Scientist and leader of the „Statistical Analysis and Modelling" group at the Max Planck-Institute for Meteorology. His research interests are climate diagnostics and statistical climatology, specifically detection and attribution of anthropogenic climate change, variability and change in storminess and related marine variables (storm surges, ocean waves), regional climate change; use of paleo proxy data to study climate variability and change, transdisciplinary context. He has published eleven books, and numerous articles and is in charge of a number of projects. He is member of the advisory boards of Journal of Climate, Meteorologische Zeitschrift, Annals of Geophysics, and organizor of the GKSS School on Environmental Research. He was a lead author of the Third Assessment of the IPCC.

*Dalia Streimikienė* – Associate Professor, Kaunas Faculty of Humanities, Vilnius University. A graduate of Kaunas University of Technology (1985). PhD (Social Science) Vilnius Gediminas Technical University (1997), Habilitation procedure (Social Science) Vilnius Gediminas Technical University (2005). Associate member of International Association of Energy Economics, UN expert in Environmental Performance Review Team, UN expert in Energy price subsidization, Energy and Environment expert in Scholarship Programme for energy experts supported by Nordic Council of Ministers. Executive secretary of the journal „Transformations in business and economics" (ISI Master List). Research interests: energy planning, energy policy and strategy, environmental regulation in energy sector, environmental economics, environmental policy, regional development, macroeconomic policy.

*Elena Talockaite*, Master of Environmental Science, currently consultant of Environmental Centre for Administration and Technology in Lithuania. She provides consultancy for local authorities in the fields of environmental engineering (air pollution and noise), sustainable urban development, involvement of stakeholders in the decision-making and strategic planning processes. Being well aware of the requirements of the EU environmental policy, as well as national, regional and local issues, she has 9 years experience of project management and consultancy, capacity building activities for local authorities and public; organisation, facilitation and chairing of workshops, conferences and trainings. Recently she has been involved in several projects focusing on climate change issues, especially adaptation and awareness raising aspects.

Environmental Centre for Administration and Technology (ECAT)
Lydos 4
LT-44213 Kaunas
Lithuania
email: elena@ecat.lt

*Anna Zollo* – expert of research in Unit for „Biology and Ecology for protection of environment" in Regional Centre of Competence Benecon

## Thematic Index

**Umweltbildung, Umweltkommunikation und Nachhaltigkeit**
**Environmental Education, Communication and Sustainability**

Herausgegeben von/Edited by Walter Leal Filho

Band 1 Walter Leal Filho (Hrsg.): Umweltschutz und Nachhaltigkeit an Hochschulen. Konzepte-Umsetzung. 1998.

Band 2 Walter Leal Filho / Farrukh Tahir (Eds.): Distance Education and Environmental Education. 1998.

Band 3 Walter Leal Filho (ed.): Environmental Engineering: International Perspectives. 1998.

Band 4 Walter Leal Filho / Mauri Ahlberg (Eds.): Environmental Education for Sustainability: Good Environment, Good Life. 1998.

Band 5 Walter Leal Filho (ed.): Sustainability and University Life. 1999. 2nd, revised ed. 2000.

Band 6 Wout van den Bor / Peter Holen / Arjen Wals / Walter Leal Filho (eds.): Integration Concepts of Sustainability into Education for Agriculture and Rural Development. 2000.

Band 7 Manfred Oepen / Winfried Hamacher (Eds.): Communicating the Environment. Environmental Communication for Sustainable Development. 2000.

Band 8 Walter Leal Filho (ed.): Communicating Sustainability. 2000.

Band 9 Walter Leal Filho (ed.): Environmental Careers, Environmental Employment and Environmental Training. International Approaches and Contexts. 2001.

Band 10 Rolf Jucker: Our Common Illiteracy. Education as if the Earth and People Mattered. 2002.

Band 11 Walter Leal Filho (ed.): Teaching Sustainability at Universities. Towards curriculum greening. 2002.

Band 12 Walter Leal Filho (ed.): International Experiences on Sustainability. 2002.

Band 13 Lars Wohlers (Hrsg.): Methoden informeller Umweltbildung. 2003.

Band 14 Ulisses Azeiteiro / Fernando Conçalves / Walter Leal Filho / Fernando Morgado / Mário Pereira (eds.): World Trends in Environmental Education. 2004.

Band 15 Walter Leal Filho / Arnolds Ubelis (eds.): Integrative approaches towards sustainability in the Baltic Sea Region. 2004.

Band 16 Walter Leal Filho / Michael Littledyke (eds.): International Perspectives in Environmental Education. 2004.

Band 17 Daniela Krumland: Beitrag der Medien zum politischen Erfolg. Forstwirtschaft und Naturschutz im Politikfeld Wald. 2004.

Band 18 Walter Leal Filho / Bernd Delakowitz (Hrsg.): Umweltmanagement an Hochschulen: Nachhaltigkeitsperspektiven. 2005.

Band 19 Walter Leal Filho / Volker Lüderitz / Gunther Geller (Hrsg.): Perspektiven der Ingenieurökologie in Forschung, Lehre und Praxis. 2006.

Band 20 Walter Leal Filho (ed.): Handbook of Sustainability Research. 2005.

Band 21 Walter Leal Filho / Dieter Greif / Bernd Delakowitz (eds.): Sustainable Chemistry and Biotechnology – A Contribution to Rivers Management. 2006.

Band 22 Walter Leal Filho / David Carpenter (eds.): Sustainability in the Australasian University Context. 2006.

Band 23 Walter Leal Filho / Arnolds Ubelis / Dina Berzina (eds.): Sustainable Development in the Baltic and Beyond. 2006.

Band 24 Walter Leal Filho (ed.): Innovation, Education and Communication for Sustainable Development. 2006.

Band 25 Walter Leal Filho / Mario Salomone (eds.): Innovative Approaches to Education for Sustainable Development. 2006.

Band 26 Walter Leal Filho / Franziska Mannke / Philipp Schmidt-Thomé (eds.): Information, Communication and Education on Climate Change – European Perspectives. 2007.

www.peterlang.de